North

Pūlena

Wailau Stream

Olaku 4602 ft.

Wailau Valley

Wailau Stream

Kahawaiiki

Str

Olo'upena Falls
Pu'uka'oku Falls (3250 ft.)
Wailele (falls)

Hāloku Falls

Pāpalaua Falls 1200 ft

campsite

Waidho'okalo Falls

Lēpau Point

Waiehu

continued on back endsheet

campsite

hula'ana
Keanapuka
(sea arch)

ka'a'ano

heiau

Kikipua

Pāpalaua Valley

Today I tried again to write about Moloka'i, but the sea was there beyond the open door. I picked up the fins, mask, and snorkel and went down across the sand and rocks in front of my house, then swam out past the reef. I floated face down, watching and motionless over the depths, ankles twined, arms crossed under my forehead, and dozed off into a light sleep. When I opened my eyes a dozen fish were hovering under me. Uhu, maiko, palani, manini, kihikihi. They thought I was a drifting log and found shelter there beneath the quiet hulk.

I came back and began —

Paddling My Own Canoe

Audrey Sutherland

A Kolowalu Book
The University Press of Hawaii
Honolulu

919.69
S

Manufactured in the United States of America

Library of Congress Cataloging in Publication Data

Sutherland, Audrey, 1921–
  Paddling my own canoe.

  (A Kolowalu book)
  Autobiographical.
  1. Hawaii—Description and travel—1951–
2. Sutherland, Audrey, 1921–   I. Title.
DU623.2.S87   919.69'04'4   78–16374
ISBN 0-8248-0618-2

Designed by Roger J. Eggers.

The endpaper maps were prepared by James A. Bier.

The Illustrations

Dorothy Bowles is the artist who illustrated this story. She is also my sister. On a solo paddle trip when the waves were breaking overhead, I was too busy surviving to take a photo, but Dot's drawing shows the threatening sea. When the cliffs loomed 3,000 feet straight up from my left elbow, no wide-angle camera lens could have included it all, but Dot could draw it. Goats, pools, sea caves, and tropical forests are all portrayed with accuracy and in the spirit of the book.

When we were kids, Dot was first to climb the Big Tree, a hundred-and-fifty-foot Jeffrey pine by our mountain cabin. She also preceded me to Hawai'i, living here for five years before I came. Now she is resident artist for the Riverside City Museum in California, close to her own wilderness cabin. She has flown over this Moloka'i country. Next year we'll paddle it together.

A. S.

Pronouncing Hawaiian Words

In the Hawaiian language, words look strangely repetitious because there are only twelve letters in its alphabet: the five vowels and *h, k, l, m, n, p, w.* The vowels have the Spanish sounds: *a* as in water, *e* in they, *i* in link, *o* in boat, and *u* in cruise or hula. *W* is often softened to a *v* sound. The macron over a letter (ā) gives it a longer stress. The glottal stop, marked by a reversed apostrophe, indicates a stopping sound where the voice pauses between letters, as in the English *oh-oh.* Syllables always end with a vowel. Go ahead, try it, and pronounce everything!

Contents

The Island   1

Preparing   7

The First Swim   15

Freeze-dried Wine   23

The Rain Fog Place   35

Wailau Valley   45

To Kalaupapa   53

The Canoe   65

Return to Pelekunu   83

Alone at the Shack   93

Pelekunu Plaza   105

Paddle On   125

Equipment   135

The Island

. . . eroded chasms, rain forest slopes, and a thousand waterfalls

*H*ula‘ana, in the Hawaiian language, is a place where it is necessary to swim past a cliff that blocks passage along a coast, a sheer cliff where the sea beats. I first glimpsed the sea cliffs and waterfalls of Moloka‘i while flying by, en route to other islands in the Hawaiian chain.

There were no roads, no trails, no people, no access except by sea. Looking down on it was not enough. I wanted to be there, but I couldn't afford to hire a boat. All right, I'd have to swim.

The plane window framed the head wall of one small bay, a single, thousand-foot face from the peak to the scree at the shore. Down one corner a narrow cascade etched a white line on the gray, lichened rock, and at its base grew a jungle of pandanus trees. Along the sides of the bay the ridged arms reached out to enfold it like the paws of a giant sphinx. Under water, eroded pockets and caves tunneled through the basalt. The shallows and the depths were evident in patterns of pale jade and deep cobalt.

Beyond that bay the cliffs rose even higher. The plane was flying at three thousand feet, but the top of the wall beside us disappeared into mists above.

We crossed a headland and flew low over a serrated plateau. Its top was suddenly sliced open and the gashed green walls fell aside. At the head of the cleft was another cascade, a thick foaming shaft that plunged over the brink and down to a dark pool, then twisted out to sea.

Later, I studied the topographic map which confirmed what I had seen from the air. Where the contour lines were so close together as to be almost solid brown, it meant only one thing—vertical, but between the guarding headlands at each end, the coast was indented by three deep valleys.

Was it possible to swim around the cliffs towing a floating pack, and to come ashore for camping in the valleys? I wondered, and searched the meager reference material. The *Atlas of Hawaii*, since published, explains the coast very simply: "The sea cliff on the north side of East Moloka'i, one of the highest in the world, is 2,000 to 3,600 feet high." *The Guinness Book of World Records* says, "Highest sea cliff North Molokai Hawaiian Islands, 3,600'." But in 1957 there were few descriptions.

Fewer than fifty miles separate Moloka'i from the half million people of Honolulu, but until tourist promotion labeled Moloka'i the Friendly Island, it was known as the lonely one.

Its western end is flat and dry, with pineapple acreage, scrubby thorn trees, and hot golden beaches. The eastern half is jagged mountains, and on the north side where the cliffs catch the moisture-laden clouds blown in by the trade winds, and rainfall is two hundred inches a year, there are eroded chasms, rain forest slopes, and a thousand waterfalls. Some tumble directly into the sea below. Others are blown up again and out by the fierce updrafts of the winds, and the air is heavy and salty with the mist of these upside-down falls. White sea foam swirls around a dozen black rock islands tossed offshore by a giant hand. It is the most isolated and spectacular seacoast in all Hawai'i.

Some inner wildness, there since childhood, surged up and answered that wild country and said very simply, "Yes. I'll come."

It would be possible to walk part of that coast. From the air I had seen a narrow edge of fallen boulders between some of the vertical faces and the water. Five flat ledges extended seaward a few hundred feet. For most of the way though, the cliffs dropped straight into the sea.

In the winter—if you can call it winter when the temperature drops from eighty degrees to seventy—the swells and storms move down to Hawai'i from the North Pacific. The waves which roll onto the gradual slopes of O'ahu's north shore are world famous in the sport of surfing, but along the north coast of Moloka'i there are no gradual slopes. The submarine canyons are deep; there is no land from there to Alaska, and the winter waves crash and climb fifty feet up the cliffs.

So I must go in summer. Then is the best chance for calm water, but there is no guarantee. In three hours the sea can change from a liquid lapping among the rocks to an eight-foot shore break, and the wind will rise from a breath to a forty-knot gale. The current and the winds move west, parallel to the coastline. I cannot swim against them. From the east end of the island at Hālawa Valley, I must go all the way to the peninsula of Kalaupapa where there is a trail out and also a small airstrip. West of Hālawa the rainfall increases, the cliffs become higher, more verdant and more sheer. Once started there would be no return.

Preparing

. . . climbed the steep trail out of Kalaupapa

I was inexperienced in amphibious journeys, but other preparation had been going on for a long time without my realizing that it would lead to Moloka'i. I had learned as a child the joy of being alone, and I wanted that isolated country.

My father was an agriculture teacher and newspaper editor. When I arrived as daughter number six, he took me along in humorous desperation to gardens and fields and mountains where he had thought to take a son. I remember standing on his instep, clinging to his high, laced boots, and later running along the trails six steps to his one. He died when I was five, but I kept tramping out alone, smelling and tasting, and identifying in field guides the plants he knew so well.

Summers in a wilderness were the best bequest he could have made. He had started building a cabin, and my mother took my older sisters and me back to it each year. We shingled and sawed, tore out the frozen plumbing after cold winters, and finally finished it ten years later. After carpentry jobs, my share of house chores, and hauling oak limbs for the wood stove, I went out alone to explore the ponderosa pine forest. I stalked deer at dusk and fireflies at night, ran wet and exultant in cloudbursts and thunderstorms, and climbed to the tops of young pine trees to swing them in whipping circles.

In college at UCLA I learned to swim, the most useful subject learned there, then, by increasing distances, developed a lazy rhythmic stroke. Water became my element. I delight in its color, its texture, the three dimensional freedom of movement, there where buoyancy balances body weight. I stagger in an alien world of gravity's pull when first I stand upright again on shore after an hour or two in and under water.

I married a seagoing man. He taught me bowline knots, warned of the barbed spines of the sea urchin, and drew diagrams of how a sailboat is able to tack into the wind. For three years we eked out a living with our own commercial fishing boat, operating out of San Pedro for albacore, halibut, abalone, and lobster—a hard, nasty, satisfying life. Between fishing trips I worked as a lifeguard and swimming instructor, did helmet and skin diving, and hiked several hundred miles of trail and cross country. Children, as they were born, were tucked on back or on board and taken along.

In 1952 we came to Hawai'i as the lowest rank of all newcomers, transplants from California, West Coast *haole*. But Hawai'i, and living by the sea in the small country town of Hale'iwa, have been good teachers. The water skills of the mainland were adapted to Hawai'i and its unceasing miracle of a warm ocean. One of the first scuba courses offered to the public in the Islands was by a pair of navy instructors. I finished it, then logged three hundred hours of bottom time over the next ten years.

So when the problem arose of how to see the northeast coast of Moloka'i at close range, an amphibious expedition in which I'd walk when possible, swim when necessary, and sometimes have a choice between the two, seemed to be a logical and inexpensive way to see some great country.

The children and I had been learning botanical and geographical lore of Hawai'i from my uncle, Max Carson, who was surface water engineer for the U.S. Geological Survey in the Islands from 1919 to 1954. Max was a mild appearing, gnarly man who often walked three days through the Alaka'i swamp on the island of Kaua'i with an eighty-pound load to check the rain gauge on top of Wai'ale'ale, the world's wettest piece of land, where the rainfall averages forty feet a year.

As early as 1903, proposals had been made to collect water from the great valleys of East Moloka'i in a high-level ditch and take it through the mountains in a tunnel to irrigate the dry flat west end of the island, but it was not

10

until 1919 that Geological Survey teams went into the valleys by boat to build gauges for measuring stream flow.

Max had told me some of the history of the valleys, but when he retired and left the Islands, I didn't know who might give advice about how to get around the cliffs and along the coast. In centuries past, some lone Hawaiian must have paddled it in an outrigger canoe. A hundred years ago, when the area was populated, some of the people were said to have swum around the intervening cliffs to visit each other when the seas were calm, but all the literature I could find simply said, "inaccessible except by sea."

By then the seagoing husband had gone back to California seas, but I chose to stay in Hawai'i to raise two sons and two daughters. The long process began of doing what was expedient from moment to moment for the children and for my full-time job. We were all prowling about the Islands, gathering guavas in the hills, setting lobster nets, diving and surfing. The first daughter, Noël, six years older than the others, was my staunch helper, and understood my wildcat need to go off alone for a while every six months or so.

It was in 1958 that I first tried to get into the remote valleys, flying to the leprosy colony of Kalaupapa in the single-engine daily mail plane, then hiking across the peninsula and along the shore until a cliff cut off access by land, and sea currents moving swiftly through the slot by Ōkala Island prevented swimming eastward around the cliff.

I went back the three miles to Kalaupapa village and climbed the steep switchback trail up the cliff where they now run mule trains down and up for tourists. I hitchhiked from "topside" Moloka'i out to Hālawa Valley at the east end of the island where I camped overnight to be ready for an early morning assault by sea. It was the end of June and the seas should have been calm, but I was flung ashore three times in the breaking surf with my pack before I gave up and settled for exploring Hālawa with its river and waterfalls.

Probably the feral pigs upriver had contaminated the water that I drank. I was sick and weak when I climbed the hot and dusty road out of the valley, thumbed the thirty miles to the airport, and flew back to Honolulu.

I had pried at the walls from one end and then the other, but the wilderness remained intact.

During the next four years, on mini-vacations to other islands with a son or daughter, we flew over that north coast, mother and child noses pressed alike to the plane window. The topographical map of Moloka'i that I clutched during those flights was old and not very detailed, a 1/62,500 scale issued in 1924, but it was the only one available and I went over it again and again.

I also kept talking to myself. "Look, Aud, there are other remote areas in this island chain. Why not try some of them?"

So one day I trudged up the long bare trail of Hawai'i's 13,796-foot Mauna Kea, before roads and observatories changed its outer space loneliness. The next day I climbed the north side of the other giant volcano, the 13,677-foot Mauna Loa. Those are the heights above sea level, but geologists look at the height of a mountain above its base, and Mauna Loa rises 30,000 feet above its origin on the ocean floor. It has the greatest mass of any mountain in the world.

I spent a freezing night on top, at daybreak crunched through the golden taffy lava along the sheer edge of the deep caldera, then headed down the east side on the nineteen-mile black scoria trail. All day I hiked groggily along, stopping now and then to curl up on the lava and nap.

That night, down to the tree line, I took off the pack and rested. A three-quarter moon rose, pink in the reflected light from the volcanic eruption of Kīlauea, four thousand feet below. Around me, as around a rock in a stream, flowed a vast river of air. Down from the chilled summit, sliding down the slopes of the immense bulk, east and west and south to the sea, the air came flowing through the night. The mountain and the wind and the darkness and I were one. I walked on, feet intuitively finding the trail that the conscious mind and eyes could not see.

I tried another place. Kalalau Valley had long been the epitome, the rallying cry, for the Hawaiian Trail and Mountain Club, so following their advice I flew to the island of Kaua'i, hitchhiked out to the end of the road at Ha'ena, and put on the pack once more. The trail followed the ancient route of the pre-European population, meandering in and out of four valleys along the cliffs—*na pali* in Hawaiian—and crossing a dozen streams.

At the end of the trail, on the far side of Kalalau, Dr. Bernard Wheatley had lived as a lone philosopher for six years in a vaulted, open cave, and was kind to the occasional hiker. From his "guest room," a sandy corner of the cave, it was a short walk to a waterfall for drinking water and a shower. In the other direction, I crawled through a rock tunnel, then made an easy round-trip swim into Honopū, with its gold sand beach and high rock arch. I did not swim on to the next valley of Awa'awapuhi, the valley of the "lost" tribe where bones of a pre-Hawaiian, Asian type people have been found. Perhaps they were the original *Menehune*, the small people, the overnight miracle workers of legend.

The children went along on short expeditions. One child is one problem. Two children are four problems in their interaction with each other and with their parent; three squared is nine—or perhaps the exponent also becomes three and the complications cubed make twenty-seven. Four children? The problems are astronomical. So they went along one at a time.

Noël dived with me among the scarlet slate-pencil sea urchins and chartreuse corals of Hōnaunau Bay. Jock caught his first trout and gorged on blackberries at Kōke'e on Kaua'i. Ann trudged with her knapsack through Haleakalā Crater on Maui, the promise of a candy bar luring her on to the next cabin. Nine-year-old James hunkered by the campfire near the crest of windswept Hualālai, broiling his steak over the coals, then wrote up his journal by flashlight in his sleeping bag.

There were business trips to Japan, and one October I climbed as far as I could up the ice-blue perfection of Fuji-san. Defeated at last by chilling winds, a lack of time, and a promised appointment, I came back down into the snow drifts of the quiet larch forest. Swirling clouds and gray lichens gently wrapped the weathered wood of an old Shinto temple. Sheltered there, I ate a delicately arranged lunch from an ash-wood box, feeling such a humble tenderness as in no other country. An ink-brushed scroll came to life as the outlines of mossy eaves and needled branches softly appeared and disappeared in the mist.

Now I had tried other mountains, other seas—and I went back to Moloka'i.

The First Swim

. . . launched, got hung up on a rock

In the spring of 1962 I began planning once more. I could hire a helicopter to take me into one of the valleys and then pick me up. I could hire a boat to do the same, although the last hundred yards might have to be in an inner tube with the gear lashed down, as a boat could rarely get any closer to the shore break. But I wanted to see all the valleys, not just one. I didn't have enough money for charters anyway, and I felt at ease alone in the sea or in the mountains.

To go without motor power I had two choices: I could swim the whole coast—twelve miles as the frigate bird, the 'iwa, flies, twenty miles in and out of the bays and valleys —or I could walk the narrow ledge of boulders and swim only where necessary around those cliffs that dropped sheer into the sea. I chose the latter.

One problem was keeping my gear dry. Backpacking wasn't yet a popular sport with the present proliferation of equipment. No waterproof pack sack was available, only thin bags and flimsy sheets of plastic. Finally, I scrounged a new rubber meteorological weather balloon, wrapped the camera, food, and clothing in it, rolled it inside a shower curtain, put the bundle into an army clothing bag, and lashed it all to a lightweight aluminum pack frame, with the fins, mask, and snorkel tied on the outside ready for use. Using bowline knots learned from the mariner husband, I tied a ten-foot towline from the pack to my waist, then tried the rig out in the calm ocean in front of my house, where it floated well enough.

For years the kids had laughed about Mom's lists: carpentry projects, Saturday morning jobs for the whole clan, morale building lists for Mom, what-every-kid-should-be-able-to-do-by-age-sixteen lists. Now the Moloka'i buy-take-do sheet was posted on the wall, and the pile of gear beneath it grew as items were assembled, and then shrank as every unnecessary ounce was eliminated.

17

I applied for a three-day leave from my job in July, made plane reservations, arranged for a rental car on Moloka'i and a child-sitter for home. Then I flew from Honolulu with daughter Noël, and a visiting nephew, Paul.

After the years of varied training and experience, and with all the information I could gather, I was prepared as well as I knew how, and yet I very nearly did not survive that first trip.

Noël, Paul, and I drove along the south shore road to Hālawa Valley at the east end, parked the car by the old deserted church, and waded across the river. Paul carried the pack while Noël and I scampered ahead toward the point, crisscrossing the rocky pastureland like a pair of bird dogs. She looked for new plants, while I found old rock walls on the higher land, and *'opihi*, the Hawaiian limpets, at the water's edge, prying them out of their shells with a practiced thumb to eat, raw and squirming, for mid-morning protein.

Around the headland on the north side of the valley, and after a mile more of boulder scrambling, we came to the end of the space where we could walk. Ahead, the first *hula'ana* dropped sheer and bare into the sea. We could not see around it to know how far I would need to swim before I could come ashore. From the map I estimated a mile or less, but I couldn't trust the map. It showed a trail up from Hālawa, over the ridge and down the cliff onto the first small peninsula, and I knew for sure from the last trip to Hālawa Valley that it had been erased years before by landslides and erosion.

I packed up the dry clothes I'd been wearing, put on the light nylon tank suit, fins, mask, and snorkel, and waited for a lull in the wave sequence. Paul and Noël stood back from the crashing surf and spray and took photos as I launched, got hung up on a rock, shoved off, and swam hard out through and beyond the breaking waves.

Treading water, I checked the towline around my waist and the floating pack—all intact. I looked to shore and waved to the two small figures at the base of the cliff. They headed back to the car for a day of sightseeing and the return flight to Honolulu. I turned and swam west, casual, confident, unaware of what the trip would bring.

An hour later and a mile down the coast, I swam ashore through the surf, exchanged fins for tennis shoes and boulder-hopped three miles around the two small peninsulas to the first valley and stream of Pāpalaua where I camped for the night.

Just west of the stream mouth was another cliff face. After I'd launched out in the morning, I decided to keep on swimming the three miles to the next valley, Wailau, where the landing would be more sheltered from the breaking waves. It took three hours, towing the pack— about the same time and with about the same effort as walking and carrying the gear along the rocks.

Onshore in Wailau I explored upstream, following the faint remains of the old G.S. trail, then came back and hiked west along the shore until another cliff cut off the walking space. I had allowed only four days before they expected me back at work, and I was pushing hard for distance.

The expedition began to disintegrate. I had no watch; it was later than I thought when I took to the sea once more. I saw the sun setting ahead and realized that I could not make the remaining two miles into Pelekunu Valley before dark. I swam ashore, climbed up to a tiny ledge above reach of the waves, and clung there for the night. There was no fresh water so I did not eat either, as my food was dehydrated and needed water to reconstitute it.

The rubber balloon that wrapped the pack was beginning to leak after the two days of sun and salt water. The food, the camera, the clothing were all damp from water seeping through.

Next morning I swam on to the river mouth of Pelekunu, drank and ate and pushed on, walking with the pack out along the west side of the bay to a ledge where an old

19

Geological Survey shack and a catwalk out over the water were rotting and rusting away. I followed the ledge out to the west corner of the bay and launched again where the cliff met the sea.

Now the pack was so wet and heavy that it took five strokes to tow it one body length even with the fins. I swam on, watching the cliff above the water to my left, and the scene underwater through the mask.

Around the point into Hāʻupu Bay I came ashore to rest. My timing was off; I was caught in a breaking wave and the mask and snorkel were torn away and sunk. I reconnoitered, then headed out again, straight across the bay, the shortest line route. The wet bundle was barely floating; it was like towing a dead and sinking body.

I swam two more hours, then came to shore at dusk as a rainstorm broke, again at the base of a cliff where there was no fresh water. I started bidding with myself as to how much I would pay for a can of juicy apricots, and agreed on thirty dollars as a fair price. I look back now and wonder that I didn't think to rig the plastic sheet as a rain catchment.

There was a small space at the base of the cliff where I could crawl under, out of the rain. When things get that bad the situation is either funny or tragic, so I laughed at the idiocy which had put me there—a faint laugh.

Both strength and judgment were giving out the next morning as I walked along shore to the peninsula of Kūkaʻiwaʻa. I knew it was a mile swim around it, and without the mask and snorkel it was hard to breathe in the choppy sea. The day before when I turned my head to get a breath, as often as not I got a mouthful of salt water instead.

I looked up at the wall ahead. It was only about sixty feet high. I thought that I could climb up, walk across the peninsula, and then go down the other side to avoid the long swim. I know better now. I should have known better then, but I had never before experienced the mind-robbing effects of dehydration.

Wearing the pack, I started up, finding cracks and knobs for fingers and toes. Ten feet from the top I ran out of handholds. I knew I could not go back down. There are no eyes in toes to help seek out the route which is visible going up.

Fifty feet below my four-inch perch on the cliff face, a rocky ledge jutted out into the ocean's white surge. Cautiously, I squirmed out of the pack and flung it out into space. It did not quite clear the ledge, but hit with a thud and broke open, spilling into the sea. The pieces of equipment floated momentarily. Desperately, I coiled and sprang outward. I plummeted into the water a few inches beyond the ledge, gathered the things that still floated and tossed them up onto the shelf, then climbed up and blacked out. I don't know how long I lay there unconscious, but I came back to reality, rested, repacked, and swam around the point.

I crawled up on the narrow shore at the base of the cliff. A noddy tern stood quietly on a rock, wings folded. "Sutherland," he seemed to say, "you're not going to make it." "Oh yes, I am," I told him, and then began to sort out the gear for the last time, discarding down to shirt, pants, shoes, the can of exposed film, the plane ticket, and the health department permit to pass through the leprosy colony of Kalaupapa where I would catch the plane out. The rest I bundled up to leave behind. I walked to the edge, ready to swim once more. It was only two more miles to Kalaupapa.

Around the distant point a boat appeared. I stumbled back, undid the ragged plastic sheet and waved it. The forty-foot cruiser edged in around the tall rock shaft of Huelo, while the crew lowered a skiff. I rolled up my gear and swam out to meet it, finning along on my back with the dripping pack held out of the water at my chin. The boatman deftly spun the skiff, stern to me, and grabbed the pack.

"What the *hell* are you doing here?" he roared. Mine

was a very meek reply, something about just swimming along the coast. Dave Nottage, who knows all the island waters from years of cruising and gamefishing, introduced himself, and rowed us back to his boat, his muscular wrists skillfully feathering the oars. Aboard, they poured me three fingers of bourbon and three cups of water, a sure cure for dehydration, then circled back to drop me off in the water near the old landing at Kalawao where I could scramble up the wooden ladder, cross to the tiny airstrip of Kalaupapa and fly back to Honolulu.

It was two years before I went back.

People said I was crazy to do a thing like that. Driving to work, scrubbing floors at midnight, nailing tar paper on the roof of my old beach house, I lectured myself, too. "Listen Sutherland, if you need the power of the mountains, the peace of solitude, the purity of water to renew your strength from time to time, why not go someplace where you know you can get back. No one else can be expected to care for the clan if you leap permanently off a cliff."

But the children figured I'd get back. They thought the expeditions were a fine idea. Maybe they decided letting me be crazy gave them more freedom.

And Moloka'i was still there. Unspoiled. I had seen reminders of the Hawaiians who once lived in the valleys, but the rocks of a taro terrace or a house foundation blended well into the jungle. A few scraps of recently shattered boats had been flung up onto the shore, a fisherman's tent was hunched beside a river, a crumbling shack and a twisted catwalk clung to a ledge, but in all the twenty miles there were no beer cans, no cigarette butts, no trash, no people.

There was a thousand-foot waterfall in Pāpalaua, a sea tunnel through a cliff, a cave on a ledge. I had seen them, I hadn't yet been in them. Now I knew what was there, but I had still barely touched that lonely coast.

Freeze-dried Wine

. . . *strange crashing sounds reverberated from the cliff*

It was the end of July 1964, two years since the first trip. An old DC 3 made the early flight from Honolulu to the main Moloka'i airport. From there I hitched a ride out along the south shore road to the Pu'u o Hoku Ranch high on the grassy plateau of the east end. Standing there at the edge of the dusty road, I looked out past the rolling green pastures with the Charolais and Hereford and Angus cattle, down to the indigo blue of the sea and across the channel south to Maui.

Suddenly there was enough time and space. Living had been such a juggling process, ten heavy balls all in the air at once. There were the individual needs and interactions of four children, the job, the housework, the car and yard maintenance, money management, a little personal life, the university classes. Sometimes one ball would fall with a loud crash. I'd lay the rest aside, pick it up, mold it carefully back into shape and toss them all into orbit again.

But now they were gone, the load was lifted. There was only the sloping green, and then the blue, on and on until the sea blended into the sky with no line of horizon demarcation, only the blue going lighter and higher until it came full circle back to me there on the hill.

Gordon MacKenzie, the tough and knowledgeable ranch manager, drove me in his jeep down the winding route to the floor of Hālawa Valley. He was inspecting roads and evaluating fields for grassland. Soon the ranch would be taking paying guests, flown by small plane from the resort areas of Maui across the channel, or from Honolulu. He drove across the wooden bridge, later washed out by floods, and stopped the jeep. He hoisted out my gear, wished me well, then turned and left. For seven days there would be no other sight of humans.

I repacked the box of equipment, lashed it to the pack frame, and lifted it awkwardly onto a high flat rock so that I could stoop in front of it, slip my arms into the straps, then lean forward and use leg muscles to rise. A luncheon knapsack can be tossed up onto shoulders, but I do not shrug lightly into a forty-pound pack, one-third of my body weight.

This time, instead of a pack wrapped in a rubber balloon, I had a two-foot square Styrofoam box in which a Royal typewriter had been shipped to Hawai'i. I had carved out the center braces and had painted the outside—one half blue to inconspicuously match the sea, and the other side red in case I wanted to be rescued. Inside I had packed the gear in plastic bags and then into a neoprene coated GI clothing bag. Lashed to the aluminum pack frame, the whole rig floated well, even when fully loaded, and rarely leaked more than a tablespoon of water in between the lid and lower half.

I had a new waterproof camera and a secondhand one with a delayed-action shutter to replace the one that died from salt and rust after the last trip.

I knew only too well now what lay ahead. There were two choices: I could use the semicalm of the bay to launch out through the breakers and swim the two miles around the northeast point, or I could backpack along the rocky shore to the place where I would have to take to the water. It would be as easy to swim as to walk, but I did not know how waterproof the foam box would remain for seven days, so I chose to walk where possible.

As I hiked out toward the point I passed the rambling stone walls and terraces of the Pāpā *heiau*, a temple site of the old religion with its many gods. Stones in ancient Hawai'i were used to fence off enclosures, to build large platforms, or to make the foundations for the living and sleeping spaces which were then walled and roofed with lashed wooden poles and thatch. I didn't know how to differentiate between all the rock remains. Which were old

26

terrace walls of taro patches, which were house walls, and which were only recent cattle pens for the ranch?

Out at the point I clung to finger- and toeholds well above the surf break, inching around Lamaloa Head to where I could walk again. Walk? This is no country for the jaunty swinging stride. The rocks are five-ton basalt boulders, jagged and upended, newly fallen from the cliffs above. I clambered up and eased down, using hands and feet in apelike progress.

The ten-o'clock sun was hot. I curled up in the narrow shade of an overhang, contouring my body to the shape of the rocks, and my cheek to a smooth slab. I slept an hour, then ate a small can of apricots, brought along in remembrance of the craving on the last trip. There were only two cans in the pack. Weight was the deciding factor for each item. I had to balance utility against ounces, substitute, eliminate, eat the heavy items first. I took the can along to use as a cup.

It was time to go again, for a mile of this coastline takes two hours. The route along the twenty-foot space between sea and cliff must be selected. Every step is on a separate rock and each must be judged or tested. Is that one solid or will it roll and turn under my weight? I used to run and leap in the meadows and from rock to rock in the streams of the San Bernardino mountains, selecting in midair the solid point to touch toe and leap again. No longer as agile, weighted by the heavy pack, and alone, I slowed the headlong pace, knowing that I would have to splint my own broken leg, and that it would be seven days before I was due back and anyone even started looking for me.

I came now to the end of the space where I could walk. Ahead the *hula'ana* cut off the narrow edge. It was only a hundred feet before the rocky shore began again beyond the precipice, but in the pounding surf there was no way to inch around the face of the cliff, floating and clinging to handholds. I would have to swim out beyond the breakers, pass the cliff face, and then return to shore.

I gave myself instructions. Take off the jeans, the shirt, the shoes and socks. Pack them inside the bag and box to keep them dry to wear for sleeping that night. Cinch the box onto the pack frame again. Put on the nylon tank suit, the fins. Lash a line to the mask and snorkel from a shoulder strap. If I get smashed by the breakers, the fins and mask could be torn off. The fins float and can sometimes be found; the mask would sink. You'd think I would have learned that lesson thoroughly. Some day I will. Tie a ten-foot towline from around my waist to the frame of the backpack.

The surf was breaking steeply, eight to ten feet high, worse than two years before when Noël and Paul came this far and took pictures as I waited for the pause between waves and then launched out.

Hawaiians put a leaf on a rock and weight it down with a small stone before crossing a stream as a symbolic offering to the gods of the stream and valley. I did the same here and then made a quiet plea to the Hawaiian god of the seas. "Kanaloa, calm thy waters." Often I have felt aware of him, but today he did not hear. Perhaps here in these mountainous islands he has less power than he has in the other parts of Polynesia where he is known as Tangaroa. There he can wash across the lower atolls and wipe out the works of the land gods.

I pulled the mask down on my face, clamped my teeth on the mouthpiece of the snorkel, and waited for a lull, standing knee-deep in the water among the sharp rocks, holding the pack on one thigh, buffeted by the surge.

Now GO!

A grunting, supreme effort. Kick, kick, swing left, don't slam into that rock jutting upward. Is the pack following? Did it tear loose? It jerks taut. A wave steepens ahead. Stroke, kick, gasp. Ah, beyond the waves. I tread water, breathe deep and fast, then start the easy crawl westward —blue side of the foam box turned up.

The pack followed smoothly along. The sea was choppy, but the water was clear and visibility good. Twenty feet below, a school of *manini* and *maiko* was feeding, so thick I could not see the rocks and growths they nibbled. There were five hundred or more, light and dark bodies packed together. The pale *manini* with its vertical dark stripes is sometimes called the convict fish, and the Hawaiian name also means stingy. Both *maiko* and *manini*, sweet to eat but bony, are surgeon fish of the genus *Acanthurus*. If you grasp them carelessly, they can perform swift surgery with the scalpel that extends on each side at the base of the tail.

I watched the school wheel and turn as one. They were aware of my presence, but had no fear, which was true of fish all along the way. If I moved slowly, they did not flee on sight as did those of Oʻahu, pursued too often by hordes of beginning spear fishermen, who didn't need the food and sought only a target.

I cruised another hundred yards with the scene below constantly changing. Over there was a *humuhumunukunukuapuaʻa*, and for once the scientific name, *Rhinecanthus rectangulus*, was just as easy. "Rather brainless-looking fishes" —a rare anthropomorphic judgment—wrote Bill Gosline, who coauthored the definitive book, *Hawaiian Fishes*.

Always there is the problem of nomenclature, and in Hawaiʻi four names are needed for every biological item. You say octopus to a mainlander, *tako* to a Japanese, *Polypus* to a marine biologist, and *heʻe* or *pū-loa* to a Hawaiian. Then you add "squeed" for the common denominator of pidgin English.

Now past the cliff face, I waited offshore just seaward of the breakers for a lull. Where was I in the continuous sequence of larger and smaller waves? It is harder to judge from the back side of the waves where your view is only four inches above sea level.

Give the mental orders: Wait. Watch. Now, swim. . . . You're sloppy; coordinate!

A wave lifts me and drops forward.

Put a hand over the face mask in case you hit a rock, unseen in the thick foam. Let the body go limp as you smash onto the shore. Grab the pack. Shield it with the body from the rocks as the surge tumbles you. Crawl, stagger. Get up to a higher level beyond reach of the next wave forming behind you. Not bad. A few slices, a few bruises, the first of many. It is inevitable.

There are only two beaches on this coast where landing is easier. At Wailau, the center valley, there is sand in summertime. In the next valley of Pelekunu the shore is usually only rocks, but one corner of it is somewhat sheltered and sloping, and at least the rocks are rounded.

I opened the box and put on the old tennis shoes, then lashed to the pack the essential finsmaskandsnorkel. It became a unit, one word. Maybe I could use FIMS as an acronym. I began again the careful stepping over the rocks, worn smooth here from the ceaseless pounding.

I climbed over a mass of rock which had fallen from the height above and hid the view ahead. Ah, there was the first of the five small coastline lava peninsulas, *lau pahoehoe* in Hawaiian. How good it looked with trees and grass in contrast to the last barren mile. It is called Puahaunui on the topographical maps, although the old spelling probably was Pua'aha'unui, reverse apostrophes being used to indicate a glottal stop where the voice pauses between syllables. The missionary reports of 1834 and 1848 spelled it Puahonui.

It is always difficult for a beginner to translate meanings. Even the Hawaiian language authorities go to the old people and find the history and legends of a place before making a definition or spelling. The twelve letters of the Hawaiian alphabet themselves were arbitrarily selected. *Pua* can mean blossom; *hau* is a hibiscus related tree which, untended, forms a tangled jungle. *Nui* means large or many. So perhaps this is the ledge where there were many blossoms of the *hau* tree. The old spelling would mean pig snorting much. At any rate Puahaunui is an area for which there are few and vague archeological descriptions.

Suddenly, strange crashing sounds reverberated from the cliff to my left. Two huge male goats were fighting, leaping up vertically and pounding down with hooves and horns, one onto the other. To one side was a herd of females and kids. I dropped down and crawled on belly and elbows toward them, keeping out of their sight behind the boulders. I had no telephoto lens, so I squirmed to the last rock, then stood and snapped the picture from ten feet away. They ignored me completely and, still fighting, moved up and across the talus slope.

I came back to my own problems. Was there fresh water here? The topo map indicated a stream, but I saw none. I might have to hike on another two miles to the first big valley stream, make camp there, and return the next day for exploration and mapping. I'd seen rock walls on the previous trip and had brought along graph paper this time to draw them for the archeologists I knew. Water was essential for all my dried rations and for me. Swimming in the sun and salt water seemed to draw moisture out of the body, and I became as thirsty as by heavy sweating. I had learned the hard way how easy it was to push too long and then quickly lose both strength and judgment as I became dehydrated and undernourished.

Rounding a slight bend, I looked ahead and saw high on the cliff the white streak of a waterfall. Good. It would be possible to camp here. I plodded on through the *hala*, or pandanus trees, taking note of the stone walls, shell middens, and open areas to chart the next day. According to the map, the stream should flow down this small gulch into the sea, but the ravine was dry and the jungle of trees impenetrable a few yards inland. I searched along the shoreline and found a grassy area, flat and dry, and reasonably free of rocks.

I undid the pack and put on jeans and shirt for protection against the scratchy lantana. Casually imported as a hedge flower years ago, and having no natural enemies in Hawai'i, this thorny shrub has become one of the worst pest plants, crowding both native forests and grassy pastures.

The waterfall was up and across the jumbled scree. I took a plastic water bag and climbed for forty-five minutes before reaching the base of the falls. A tiny pool, surrounded by ferns, overflowed with small liquid noises and ran down the gulch; then its water was absorbed into a tangle of underbrush.

I looked up. Far above, the water spurted out from the cliff and tumbled down, touching and leaping out again. As clouds moved back across the top of the cliff, the whole wall seemed to tilt toward me. My throat caught in fear, but the falling wall was only an illusion, often repeated when I walked at the base of the cliffs. Later I checked the contour lines on the topo map. Twelve hundred fifty at the top and two hundred fifty at the bottom. An airy thousand-foot fall and many more to come.

With the water bag filled for supper and breakfast, I started back down, but there were birds to watch and the sea to scan, so I did not hurry. The coastline to the west was spectacular and ominous. The end of the journey, the peninsula of Kalaupapa and its airstrip, was only a misty blue line on the horizon.

Back in the grass I set up camp, with each item taking on double duty. The nylon cord that had lashed the foam box to the pack frame was now strung taut between two trees. A plastic sheet, which had served as waterproof wrap and padding, hung across the cord, its edges held out by rocks to make a tent. I scooped small holes in the ground between the rocks to form plastic pockets for catching rainwater during the night to increase my drinking supply for the next day. I put the candle and the flashlight inside the tent, made a bed of grass and leaves, gathered firewood, and chose a supper bag.

I had packed each meal complete and separate, and had inserted the menu for each one on a slip of paper inside the see-through plastic bag. Part of the idiot fun of these expeditions has been the reverse twist, creating sybaritic luxuries where all should be hardship, and transferring the gusto of a gourmet kitchen into meals that weigh an average of twelve ounces per day. Eight days would require six pounds, with the addition of some food from the land and sea, caught, picked, or scraped.

That first night I leaned back against a black rock, still warm from the day's sun, watched the sunset, and dined on thick beef soup with tiny dumplings, an excellent canned French Beaujolais, and juicy strawberries reconstituted from freeze-dried. My son James often used to moan, "Hell, Mom's going off on another expedition. She's taken all the strawberries out of the cornflakes." The total pack weight of supper had been ten ounces. The remainder of the can of wine I poured into 35-mm film cans, which held an ounce each, to provide an epicurean touch to later meals.

The canned wine was evidently an experiment that didn't work out. I've never since found wine in cans, although they're better than bottles for backpackers. Plastic envelopes would be a further improvement. Though we now carry out everything nonburnable, the wilderness ethic then was to bury trash, and buried somewhere under Puahaunui soil, a French wine can lies rusting away.

What I really need is for some scientist to develop a dehydrated or freeze-dried wine. Please forgive such sacrilege, Monsieur Lichine and Mr. Balzer and you other connoisseurs, but I do enjoy wine with my meals, and seven half-bottles, a week's supply, weigh ten pack-sagging pounds. Table wines are twelve percent alcohol and perhaps two percent grape residue. Perfect a dehydration method and I could carry a fifth of that lovely wine, Louis Martini's Moscato Amabile, in a container holding four ounces. Develop further; freeze-dry the alcohol. Then I could buy foil packets of a powdered Beaulieu Cabernet Sauvignon, or, for Franco-oenophiles, a Chateau La Mission Haut Brion, add water, display the packet label with a flourish, and pour with a drip-stopping wrist twist—into a Sierra Club cup. "But listen, Aud," say my scientific friends. "If you really want concentrated wine, it's already been done. It's called brandy."

The sky was cloudy and the annual average rainfall along this coast is two hundred inches; so I rigged for rain, then went to bed by candlelight and wrote up the journal for the day on the back of the map. Below the clouds the North Star was clear in the sky, twenty degrees above the horizon. To the west was an intermittent flare of light. I watched until the regular ten-second interval became apparent and then knew what it was. The Japanese poetry form of *haiku* came to mind with its own formal rules and delicate charm. Three lines: five syllables in the first line, then seven, then five. I juggled phrases.

> From Kalaupapa
> Lighthouse beams its flashing ray
> On a far small tent

I slept. During the night it rained, but I stayed warm and dry. A wild pig wandered by. I heard him snort and flap his ears, and felt his footfalls. So perhaps it was Pua'aha'unui, pig-snorting-much. A young goat was crying on the cliff. Always one circuit remained tuned to the noise of the sea, still loud with surf that had not calmed. The last quarter-moon rose and woke me with its light. I slept again.

The Rain Fog Place

. . . hunkered by the flames, sipping tea with rum

Daybreak came with a steady rain, so I stayed in the tent and watched the birds. Propped on my elbows under the streaming plastic roof and looking out, I followed the flights of *koa'e'ula*, the red-tailed tropic bird, and then the terns wheeling and soaring from their ledges in the cliff. Although water animals appeal to me more, some zoologists say that ninety percent of the study and money going into animal research is for birds. Is it man's dreamlike yearning for flight, or his atavistic remembrance of the time when his ancestors were more closely related to the pterodactyls, that creates the fascination for birds?

The dawn rain stopped, and I spent the morning mapping the remains of the old Hawaiian sites. There were dozens of rock walls; some appeared to be foundations for houses whose frame and thatch had long since rotted away. *Pū hala*, pandanus, had covered many remains. I floundered in the dark gloomy shade of the jungle, dropping now and then four feet into a hole, unseen under the leaves whose serrated and spiny edges were shredding my ankles and arms.

One wall seemed to enclose a *heiau*, and its area was open and clear in contrast to the tangle around it. Nearly two hundred feet long and a hundred feet wide, it had only three *hala* trees in the clearing. Could a *kahuna*, a Hawaiian priest, have forbidden the trees to encroach? I know only that strange, lovely things have happened in the lonely places of Hawai'i, and I am willing to believe.

Are there remains of a missionary church here under the leaves? In May of 1848, Mr. C. B. Andrews, a missionary stationed on the south side of Moloka'i, wrote:

> During the last year meeting houses have been built in Kalaupapa, Wailau, Pelekuna *[sic]*, and Puahonui. . . . In Pelekuna and Wailau, two deep dismal vallies with only

about 100 or 150 inhabitants each, by dint of hard labor, they (the missionaries) each have a house . . . and the cleanliness and pleasantness of these sanctuaries of God, compared with the miserable habitations of the people, allure many to public worship who otherwise would probably not attend. Following on in the same direction on the Kolau [sic] (east) side you come to Puahonui with its fifty inhabitants and neatly plastered stone meeting house.

Where are the remains? The dark stone walls, unmortared, of the Hawaiian hut foundations and the *heiau* still stand through crumbling effects of rain and lichen and tsunamis, but where are the walls of the limestone cemented and plastered meetinghouse?

Late in the afternoon I packed up, erased the campsite, and moved on, winding around the shore of another ledge, Hāka'a'ano, toward the first of the three north coast valleys, Pāpalaua. I trudged over the boulders, then set up a self-portrait. I screwed the camera with the delayed-action shutter to the clamp tripod, fastened the clamp to a driftwood branch, and braced the branch upright in the rocks. I checked the light meter for proper exposure, set the aperture, and sighted the camera onto the area ahead, making sure to include the waterfall in the shadowed mist at the head of Pāpalaua Valley. I poised, pushed the timer release, then skittered with the backpack over the slippery rocks to get into the sights of the camera, as the shutter would click in only seven seconds. "Look casual, Aud, like just hiking along." There are very few self-portraits.

The last quarter-mile that day was half swimming and half wading under the huge arch of Keanapuka. "The hole cave" is an eroded cavern punched through a tongue of rock a hundred feet high and extending two hundred feet into the sea. The arch beneath echoes with the grumble of boulders rolling in the surf. The sun was down and the wind still gusted along the shore. Ahead the clouds came swirling down into the darkening valley, hiding the waterfall.

Wet and cold I stumbled on, came to the mouth of Pāpalaua Valley and searched for the cave which is marked on the topo map. It had disappeared, probably during one of the frequent cloudbursts when the swollen stream had cut back its banks, and the cave roof, no longer supported, had collapsed. There was no natural shelter. At dusk I rigged a driftwood and plastic lean-to up in the *hau* and *hala* forest, away from the windy shore, then built a fire to cook supper and dry the clothing.

Dinner was an herb rice casserole with dried mushrooms, wine, cheese, and fruit. I peeled down to the high topped tennis shoes and clumped off to the river with the dirty dishes. Alone and content among the trees at the water's edge, I stood like Daphne, bewitched there in the forest. Daphne, ha! Where's Apollo, you dirty, salty female? I knelt by the pool and scrubbed, composing a derisive *haiku*, as did Basho and Issa in Japan long ago.

> Goddess by the stream
> Tall, bare, proud . . . laughs at dreams, and
> Squats to wash the pots.

I rinsed the dishes and myself in the cold pool, then wet and shivering, climbed back to dry by the fire. The surf still crashed and roared, louder than the stream's higher pitch. When would it subside, that sound, rolling into the valley, penetrating the jungle? The *hala* trees seemed strangely human in the firelight, their finger leaves spread at the end of bare angular branches, and their trunks reaching down and out with fanned toes to support the bare shafts.

And now, this second night out, hunkered there by the flames, sipping tea with rum, feeling its warmth inside, the fire's warmth on my face and shins, and the wind's chill on my bare back, I felt again the surge of pure primitive joy and power that comes with being alone and wary and confident. I had known it at age fourteen when in adolescent rebellion I took a blanket, raisins, and cheese, and walked for three days into California's San Gorgonio Wilderness Area, seeing no one and reveling in the solitude. I know it often in and under the sea, in that blue world where I can move with a dancer's leaps and spirals, a seaweed's supple sway, no longer an awkward stick of a land animal. I squatted there, hugging my knees and grinning.

The clothes were dry and I pulled them on. I carried no sleeping bag or air mattress on this trip. Their weight and bulk overruled their comfort. Sleeping in jacket, jeans, and socks, with legs stuffed into the clothing bag, was adequate. At least there were no mosquitoes near the shore in these valleys despite the high rainfall.

An inveterate reader in bed, I took a battered paperback of John Muir's *Mountains of California* out of its plastic bag, rolled under the lean-to, and read by candlelight for my allotted half hour. I had measured the burning rate of various candles. This one consumed one inch per hour. For the trip's seven nights, I had packed a three-and-a-half-inch stub to use for reading and journal writing at bedtime.

At daylight I awoke, a bit creaky from the bruising on the rocks the day before and from sleeping on the hard ground. A brown four-inch spider was crawling across the inside of the clear plastic sheet that covered the lean-to. Outlined against the light it moved along, three inches from my face, yet seemingly unaware of me. Using the left front leg almost as an antenna it reached and probed, questioning this plastic object. I questioned too. Did the eight-legged arachnids evolve first, or the insects with their six legs and two antennae?

I rolled out of bed and breakfasted on an ounce-and-a-half quick utility meal—instant oatmeal, coffee strong enough to stiffen up the spine, and a vitamin C tablet—then started off to explore the valley.

The sun was bright on the upper rim of each side, but the valley was so narrow and the sides so steep that there was sun on the valley floor only four hours a day. At ten it rose up over the east rim, and by two it had set in the west, but the sun often does not shine there at all; the very name Pāpalaua means rain fog. *"Pupuhi kukui o Pāpalaua he'ino,* light the lamps of Pāpalaua, the weather is bad," said the old Hawaiians throughout the Islands on gloomy days.

My progress was slow through the jumble of vegetation along the sides of the stream. Wild goats, black, tan, white, and every combination, were plentiful. I was quietly watching two of them feed when behind me I heard a desperate high-pitched "maaaa." A tiny kid stood there, cut off from its mother by a huge new upright beast it had never seen before. I photographed him, a small black tautness there in the green world, then stood still while he moved away through the trees to join the others.

Goat kids are charmers, but they grow up and propagate, and goats as a whole are an ecological menace in Hawai'i. Carried on board the ships of Captain Cook and Captain Vancouver in the late eighteenth century to provide a continuous source of fresh meat for ships' crews, the goats escaped their pens, went wild, and multiplied to become a scourge. They eat the native plants down to the bare earth, the soil erodes, and the symbiotic relationships between endemic Hawaiian plants are destroyed. Foreign plants, imported accidentally or irresponsibly, move in and crowd out what remains of a native flora that exists nowhere else on earth. The birds which evolved over thousands of years, uniquely adapted to feed on native plants, die also. More bird species have become extinct in Hawai'i than on all continents of the world combined.

As I pushed on, the growth became nearly impenetrable, although the rock walls of the taro terraces were still visible. When and how did the Hawaiians live here? The first migrations to Hawai'i probably came north from the Marquesas Islands about A.D. 500. Then around 1200 the second migration came, possibly dispersing from Raiatea, Tahiti, and Bora Bora. A more recent theory offered by Dr. Yosihiko Sinoto of Honolulu's Bishop Museum, holds that the Marquesas were the dispersal center for both migrations. The second group was more warlike, and more rigid in their social structure. Class strata were strictly enforced with tabu laws (*kapu* in Hawaiian), and human sacrifice was not uncommon.

In these valleys of north Moloka'i, I often felt that some of the ancient people were squatting with me by the fire. Were they those of the first migration pushed back to this rugged wet isolation by the later fiercer groups who moved in to occupy the flat, sunnier southern shore? Hawaiian was not a written language then. The only record of the thousand years is what survived in verbally memorized chants and legends to be written down after the arrival of Europeans and Americans. Surviving too, is the spoken language with its rich historical background, its idioms, its imagery and metaphors.

The sides of the valley steepened so that it was even more difficult sliding and clutching through the brush. I moved to the stream, the only clear area, and gave up any pretense of staying dry. I shifted into four-wheel drive— the use of hands and feet—splashed into the water, and simply climbed up through the waterfalls. Finally the gradient flattened, the stream widened and slowed.

Fragrant yellow ginger blooms hung over the deep pools. Filigree shadows of '*ōhi'a* leaves moved lightly over the flat rocks, while overhead the climbing pandanus, the '*ie'ie*, twisted around the convoluted branches of *kukui*, the candlenut tree. *Ti* plants, nibbled by goats, grew in shiny clumps on one side of the river, while on the other, the dark cliff rose, with bouquets of fern and moss clinging to crevices. Tiny mists of dripping water sparkled on the soaring wall in minute prisms. The green was overwhelming, emerald and olive, chartreuse and jade, shadowed and golden.

In a trance, I drifted around the last bend. High above on the back wall of the valley the thousand-foot drop of

Pāpalaua Falls disappeared into a chasm. The hidden white shaft spiraled, reappeared, turned down again through a cleft, then hurled out for the final eighty-foot roaring drop into the clear dark pool at my feet. My conscious mind returned. My fingers hurt. I had been standing with arms stiffly out at my sides, fingers spread, rigid and mesmerized. I did not swim in the pool. It glittered there, bottomless and black, a place for lurking *mo'o*, the evil water spirits.

I slipped and scrambled back down the streambed to camp, then sorted and packed the gear, ready for takeoff the next morning. I lay there that night, looking out at the blackness, listening to the stream and the sea and the closer sounds of twigs snapping, and thinking of those *mo'o*.

In 1915, Nathaniel B. Emerson, Jr., wrote of this area and of Hi'iaka, younger sister of Pele, the goddess of volcanoes. It was one of the legends so carefully memorized by each generation and passed on to the next as a heritage.

The good people of Halawa Valley, where Hi'iaka found herself well received, made earnest protest against the madness of her determination to make her way along the precipitous coast wall that formed Moloka'i's windward rampart. The route, they said, was impassable. Its overhanging cliffs . . . dropped the plummet straight into the boiling ocean. Equally to be dreaded was a nest of demon-like creatures, *mo'o*, that infested the region and had their headquarters at Kikipua, which gave name to the chief *mo'o*.

Kikipua, being of the female sex, generally chose the form of a woman as a disguise to her character which combined the fierceness and blood-thirstiness of the serpent with the shifty resources of witchcraft, thus enabling her to assume a great variety of physical shapes. . . .

The way chosen by Hi'iaka led along the precipitous face of the mountain by a trail that offered at the best only a precarious foothold or clutch for the hand. At one place a clean break opened sheer and straight into the boiling sea. As they contemplated this impasse, a plank, narrow and tenuous, seemed to bridge the abyss. Wahineomaʻo (Hiʻiaka's attendant) promptly essayed to set foot upon it. Hiʻiaka held her back, and on the instant the bridgelike structure vanished. It was the tongue of the *moʻo* thrust out in imitation of a plank, a device to lure Hiʻiaka and her companion to their destruction.

Hiʻiaka, not to be outdone as a wonder worker, spanned the abyss by stretching across it her own magical *paʻu* (skirt), and over this as on a bridge, she and Wahineomaʻo passed in security.

The *moʻo*, Kikipua, took flight and hid among the cavernous rocks. . . . Hiʻiaka gave chase and put an end to the life of the miserable creature. Thus did Hiʻiaka take another step towards ridding the land of the *moʻo*.

I fell asleep and dreamed of fierce, writhing demons. It was to Kikipua peninsula that I would be going in the morning.

Wailau Valley

The waterfall . . . filled the space between the cliff and sea

Between Pāpalaua Valley and the peninsula of Kikipua, just west of the stream mouth, was another short cliff face to swim around. I watched for the place among the rocks where there was a hint of a channel or fewer big rocks awash in the surge. Was there a rip going seaward where an outgoing current flattened the incoming waves? Good. I could use it to hitch a ride out.

Wade out knee-deep, wait for a lull in the wave sequence, then lie out instantly to horizontal. Kick hard and deep.

The head points the way, the hands are pectoral fins for balance stabilizers and for the small quick turns. The soft and buoyant sea is easy to get into; it's the getting out that tears me up. Perhaps the claws of the dead *mo'o* are still there, upturned among the rocks.

Once again on shore, I moved on to Kikipua, the third of the flattened ledges along the coast. There are no land routes to it from the rest of the island; like the eastern two, it is accessible only from the sea.

On these first trips into north Moloka'i, there was no compilation of archeological information to read before I went. It was not until 1971 that Catherine C. Summers' book, *Moloka'i: A Site Survey*, was published by the Bishop Museum. Mrs. Summers credits the work of seven men for most of the archeological data in her report. Among them was John F. G. Stokes, former curator of Polynesian ethnology at the museum, who wrote about the *heiau*, the temple site on the east side of Kikipua. Legends said it was built by Alapa'i, a great shark hunter of Moloka'i who captured fishermen and placed their decomposing bodies on a dish, which was then taken out to sea, the drip from the decomposition attracting sharks. Stokes thought that the Kikipua area might have been used as a training site, a college for priests because of its isolation.

Isolated it is, and overgrown. For ten months of the year it is surrounded by row on row of breaking waves. Even in summertime trips around the point, eight by sea and one by land, only once have I known the ocean to be calm enough for a boat to come ashore. Possibly a rubber raft with all gear lashed tight could bounce-land on the rocks in the surge. Nor is there a clear space for a helicopter to put down, and it would take weeks to clear the prickly *hala*. I was not even aware of the walls of a *heiau* on the east side, it was so overgrown, but out on the end of the peninsula were dozens of rock structures.

There are no Mayan treasures here, no hieroglyphics, or artifacts that lure vandals, but a trained archeologist, looking for a subject for a doctoral dissertation, and willing to be baked, shredded, and drenched, would have an untouched site.

I had no desire to become a piece of Alapaʻi's decomposing shark bait, so I walked warily, testing for holes under the *hala* leaves that might be traps, then plodded on along the narrow shore below the cliff. The topo map said 2,250 feet up, almost straight overhead. What if another of those boulders at my feet should topple from the heights now? I decided that even with Kikipua and Alapaʻi prowling around, my chances were statistically better than on a freeway, so I kept on at a steady pace, carefully balancing at each step on the rounded rocks, and well aware of the weight on my back of the foam box, now down to thirty-eight pounds. The waterfall of Waiahoʻokalo, water-for-making-taro, filled the space between the cliff and the sea. I walked through it, tilting my head back for a quick drink.

I made an estimate of what time it was from the sun's position, then pulled the tide chart out of its plastic bag in my pocket. Various ports around the Islands were listed with a plus or minus in minutes to calculate the time of high or low tide from the main chart for Honolulu harbor. There was no port listing for the rocks and cliffs of north

Moloka'i, but I estimated it as an hour earlier than Kaunakakai on the south shore. Ahead was another *hula'ana*; the tide was as low as it would get that day, so I squeezed around, sloshing through the waves and clinging to the vertical wall. Finally I switched back to swim gear and launched out for the last half mile swim around Lēpau Point to Wailau Valley.

Near the shore the sea was choppy and murky but out beyond the surf it cleared. Yards away I saw him coming toward me under water, an elongated oval on a collision course, becoming more enormous with each prolonged moment. The oval head was two feet high—high not wide—oh, not a shark. What then? With one quick flick of his tail fin he veered and slid past me, my towline, and my floating pack. The high, flat body, the hump head, were unmistakable now—*ulua*, the crevalle, a prized game fish about five feet long.

I swam on around the point to the calmer water on the lee side of Lēpau Point, then on in to shore. I hauled the gear up onto the rocks, and made camp in the shade of the jungle, away from the old tent of occasional fishermen on the shore. Back under the trees, all was cool and green with giant elephant's ear leaves as big as my arms' span, wild ginger with the throat catching fragrance of its white flowers, *mauna loa* vines with their shaggy pods hiding the polished brown seeds, and overhead the canopy of light green leaves of the *kukui* tree.

After losing part of my lunch to a marauding rat, I hung the food bag on a cord and looped it from a tree branch, then went off exploring. On the topo map Wailau Valley looks like a web spun by a drunken spider, the contour lines and the stream courses converging, crisscrossing, running parallel, then spreading out to the river mouth. Wailau is the largest of the north Moloka'i valleys, indented four miles from the sea to the back wall, and a mile wide.

Following the stream I went a mile or so up the valley.

Until 1955 a trail was kept open by the Geological Survey teams who went in periodically to check their instruments. The first stream gauging stations were installed in Waikolu, the valley nearest Kalaupapa, in June of 1919, and in the other valleys later that year, most of them built by engineer Benjamin Rush and a crew of laborers. There were a number of houses connected by pebbled paths in Wailau Valley at the time, near the beach, and although permanent residents had left, they were usually occupied in summer by people gathering *hīhī-wai*, a freshwater shellfish which lives on the bottom of the streams.

Max Carson, my uncle, had reminisced about other Wailau events. Getting in and out of the valley was never an easy matter even when the old trail existed. An unequaled feat was that of James Naki, who used to carry the mail from Wailau when there was still a settlement there. His wife became ill, and in the span of a day and a half, Mr. Naki went to Pūkoʻo on the south shore over a ridge three thousand feet high to see the doctor, returned with medicine, made the trip again for further instructions, came back to Wailau, and returned to Pūkoʻo with another report. As *ʻaukuʻu*, the heron, flies, this is seven miles each way, and every hiker knows that you have to double or triple this to get trail distance.

During the 1950s landslides, rain, and jungle growth obliterated much of the route from the top of the *pali* at the back of the valley down to the stream. A group of Hawaiian Trail and Mountain Club members hiked it in 1952. In June of 1969, five University of Hawaii students needed ten days to make the round trip.

Recently, trail clearing parties sponsored by the Sierra Club and the Hawaii State Division of Forestry have worked for four summers in the mud and rain, restoring the trail from the sea to the back of the valley.

Because of the trail building, because Wailau's geography is more inviting than the other valleys, and because there is a precarious but slightly sheltered anchorage in the lee of Lēpau Point, the valley is the most often visited of those along the coast—about a hundred people a year.

As I followed the stream up the valley, I kept peering into the water looking for the small silvery fish I'd seen on the previous trip. In my pocket I had some monofilament line and fish hooks, but as always when you're well prepared—no fish. Later I learned that they were schools of fresh water mullet which come and go.

There must have been a few of them still hiding in the shallows. Coming back to camp I had to recross the main stream. As I waded out of the river where it fed into the ocean, two big gray fish swam past my legs, heading up against the flow. Small sharks? No, the dorsal fin extended too far along the back; the base of the tail was too small. I grabbed the face mask from the pack and slid deeper into the water to get a better view. As they swam away upstream into the cloudy area where fresh and salt water met, I saw that they, too, were *ulua*, two and three feet long. Freshwater *ulua*? Ah well, Hawai'i is a land of contradictions. Ironwood sinks and pumice rocks float, rainbows appear by moonlight, and waterfalls spiral upward in the wind. It seems that *ulua* often swim up the river, there to feed on the mullet, and on guavas and mountain apples washed down by rainstorms.

In Wailau too, was an old mule, gone wild from the days when the G.S. teams used them to haul gear from shore to their camp upstream. It was last seen by a hunter who had gone deep into the valley with bow and arrow. But he searched for wild pigs not mules, and the old one may still be there, his sunset hee-haw echoing around the lonely valley.

There are other old remnants. After supper I prowled back into the jungle looking for house sites or pebbled paths. Against a small cliff only a few yards inland there was a rock jutting out, forming a low wall. Around it was looped an old chain with rusty, flaking links each six inches long. The end to the right of the rock was broken off, but the left end came straight toward the sea, taut still, then disappeared under the tangle of ginger, banana, and *hala* before it reached the rocks on the shore.

What ship did it anchor? What is the rusting rate of anchor chain? Would the shape and size of the links and analysis of the metal tell to which decade of sailing ships it belonged?

I fell asleep that night thinking of Vancouver and La Pérouse and whaling ships, visualizing a sudden storm and the surf rising, and a desperate crew chopping through the thick manila line that held them to the chain, while others hoisted sail to claw off this hostile shore, leaving the chain to slowly rust away there in the jungle.

. . . flutter kicked along, burbling sea chanteys

My subconscious mind must have been perking along all night on the vision of square-rigged ships, spars, and sailors. I stood there in the morning, coffee cup at my lips, looking out at the sea, and suddenly said aloud, "Wonder if I could rig a sail on this floating pack and let *it* tow *me?*" I found some bamboo poles among the driftwood to use for masts and a yard, an old shirt on the beach for a sail, and tied them onto the pack frame along with the foam box. Then, a walking shipyard, I trudged westward along the rocks to the black sand beach at the far end of Wailau, and on around Waiehu's flat curve.

Ahead were the highest waterfalls of the coast. A dozen or more emerge from the plateau above, and because of heavy rainfall they were superb that year. Five of them are nearly three thousand feet high, but they do not make a simple, single drop. The cliff is not quite vertical; it leans back at a seventy-degree angle. Down the green slope the white water slides and splashes, bounces outward, touches and vaults again. *Wailele*, leaping water, catches the wind and sheets across the slope, then spirals upward with the air currents. The twisted sprays reflect the light in a thousand rainbows, drift sideward again, and become part of the next plumed cascade along the precipice.

I walked along the shore past the base of the falls. Ahead now, another *hula'ana* cut off the walking space. There were more than two miles of cliff before I'd be able to come ashore again. I slid the pack off my back, got out the bag of nylon cord and began building my square-rigger. I lashed a pair of three-foot masts to the lower extension of the Himalayan pack frame, then tied a third piece of bamboo across the top of the masts as a yard.

With the shirt tied in place to the yard above and down to the foam box below, the craft was ready for christening. What could I do it with? My freeze-dried champagne was still only in the wistful dream stage, there would be more than enough seawater, and to pee was irreverent, so I spat respectfully on its name plate and proclaimed with proper oratorical tones, "I christen thee the *Royal One.*"

We launched through a slight lull in the six-foot shore break and pushed on out to deeper water, then veered to the west. It worked. Hooray! Our speed was a knot and a half, our tonnage seven-hundredths. We were too low in the water to catch much wind, but at least the *Royal* pulled itself. It worked best when I grasped the lower ends of the masts, then added the rudder and auxiliary power of my fins. I flutter kicked along, burbling sea chanteys through my teeth which were clamped on the mouthpiece of the snorkel, using the rhythm of the kick to beat out the cadence for the song. "Sai–*ling,* sai–*ling,* over the bounding main."

The depth was ten to twelve fathoms with the bottom clearly visible. In places a ridge would rise to only three fathoms, but my craft was of shallow draft, one foot, so there was no danger of going aground out there, especially with the ten-foot swells that were lifting us and passing on.

We came by the first arm of the small bay of Kaholaiki with its vertical head wall and cascade that I'd seen from the air years before, the glimpse that started all the journeys. I would have to come back to explore it; for now the next port was Pelekunu, another mile. Past the small bay I hauled down the sail, as the wind was constantly shifting. It would be another hard hour's swim in to the shore of Pelekunu Valley against the outgoing river current, and all on fin power, since I was still pushing the box ahead of me. With its bare masts and no wind it was too tippy and unstable to tow.

The water in the bay of Pelekunu was murky, a blue muddied by the dirt that the rains continually gouged out of the steep valley sides and washed into the sea. I passed a stream tumbling out of its dark green glen of foliage, splashing white down the black rim of the bay. I beat past

56

the carved sea caves, then flutter kicked on into the breakers which flung us high onto the rocky shore. I stood and untied the towline and looked around Pelekunu.

Like a giant thumb print in wet earth, the valley pressed into the island. One translation of the name is "the burning throat of Pele, volcano goddess." Probably more authoritative is the meaning, "smelly from lack of sunshine." It is. It smells of moss and decayed logs, ginger blossoms, rotting guavas, the stench of overripe *hala* fruit, and old goats.

I camped on the cobbled riverbank in the plastic tent, awakened to steady rain, slept again, then despairing of out-waiting the rain, soggily sat and repaired one of the bamboo masts, preparing for a late morning take off. A Hawaiian owl, the enchanted *pueo* of local mythology, stood on a rock across the river, his plumage ruffling in the wind and his head swiveling nearly a full circle. His eyes, unlike those of most other birds, are in the same plane, and he focused on my work out of both calm yellow eyes.

Rigged to go, I crouched in the shore break, holding the heavy box on one leg and standing sideways to the waves. The rolling rocks gouged my ankles and tore the heels out of the socks I wore for padding beneath the fins. Twice I swam hard, trying to get beyond the surf, but was washed back on shore. It was easy to dive under the waves, but the high floating pack, lashed to me by the towline, each time would be caught in the white water and tumbled to shore, towing me with it. Fearful for the foam box and the masts, I tried to keep one leg under the box as we hit sand and rocks.

Finally I gave up and decided to try from the ledges along the west side as on the previous trip. As I started up the trail leading to the tiny old Geological Survey shack, the rain broke into a downpour. I was cold and discouraged. The muddy, slippery trail angled up along the cliff. A misstep here would not mean just a bruise, but a plunge down fifty feet or more to the rocks below. On further, the path was so overgrown with sisal and guava and *hala* that only my feet could find it. My eyes found it only if I crouched low under the brush.

When I reached the shack I was tempted to go on, to try for the end goal of Waikolu that night, but the rain continued and the sea was choppy. Besides, the little cabin intrigued me. It was as I remembered from two years before. There were five bunks, the remains of a kitchen and a shower, a rough table and benches, all in a space of eight by twelve feet. Half the boards were gone from the roof, the bunks and the floor were sagging with damp rot. Part of the floor had already rotted into the wet ground beneath, the panes of glass were broken from the windows, and rat and goat droppings covered the floor.

It was a good place for solitude though. For five days now I had seen no people. Not even a boat had gone by, and the only planes were high and far away. So I decided to stay and rig everything for an early morning start the next day. Rain would not keep me abed in a tent, for now I had a stand-up shelter, such as it was.

I took four pieces of a moldy, warped plywood packing crate, a rusty hammer, and climbed the cliff in back of the shack to where I could turn and belly onto the roof, then nailed them on, shingle style. It was not a job I would have predicted for this trip, but neither were square-riggers.

In the afternoon I sat on the steps outside the sagging cabin door, and brought the journal up to date:

It's about 3:00 P.M. My wet clothes are hung on a line in the former shower stall. The rain has stopped and the wind has died, and even the sea has calmed. Was it so rough this morning so that I would stay one more day and come to this place? Ho, idiot, does the sea flow only for you? Now the sun shines dimly onto the beach through a cleft in the peaks behind me, but already it has dropped too low to be seen here. I look at all the ridges and peaks up the valley, then at the topo map. That must be Kamakou, the highest mountain of Moloka'i, rising into the clouds there at the head of the valley, and the ridge to the left is Pōhākaunoho—the resting seat—separating this valley from Wailau. Now the clouds roll down from the ridge and it rains again.

A hundred feet from the shack a little stream bounced down the rocks. The pipes which used to carry water from stream to sink and shower had long since broken, and the sink had rusted away, but the three small pools were there, as Uncle Max had described them to me ten years before, the upper one for drinking, the middle one for bathing, and the lower one for the laundry.

Between the shack and the stream, on a lower ledge, was the marvelously efficient outhouse. Two four-by-fours spanned a chasm which dropped thirty feet to the open sea below. The walls, the floor, and a sociable two-place seat with the patina of long use, were all sturdily constructed, but some of the boards of the floor had been torn away in high winter surf. There was no door, and the view across the bay was marvelous. No architect has ever placed a building in a more functional setting. Rain beat a crescendo on the tin roof, a counterpoint tympani to the surging sound of the automatic flush as I looked down between my knees and watched the sea rushing in and out of the cleft below.

Directly below the shack was a narrow catwalk extending out over the sea. It was built of steel beams and wooden planking, supported by a cement frame below and with a rope ladder swinging from the end. All the beams were rusting now, and the steel support cables were fraying from their anchored eye bolts in the cliff above the shack.

At dusk I built a small cooking fire and knelt there stirring a pot of herb rice with mushrooms. The wooden bunk in the shack was as hard a bed as the ground of previous nights, and I woke often to the sounds of rain and sea.

In the morning the *Royal* and I made the easiest launching of the week. I lowered the box-boat on a line from the catwalk, climbed down, and carrying the box, clambered along the rocky lower shelf as far as I could toward the mouth of the bay. Then with fins, mask, and snorkel on, and the boat on my lap, I sat on the edge, waiting. The surge sucked away below me; as it lifted again ten feet to my level, I fell forward into it and dropped with the water, then started swimming away, towing the box, its masts unstepped and lashed to the "deck."

I gave myself orders. Swim along the cliff as close as I dare go, keeping just beyond the white surge that leaps upward at my left. Could I climb those walls in case of danger? Probably not. Still they seemed more protective than the open sea. Look down through the glass of the face mask. Keep a steady six-beat crawl.

The angle of the cliff stayed constant above and below the surface of the sea. I measured with my mind's eye, down from coral clump to cluster of fish to jutting lava outcrop—ten, thirty, fifty feet down. The colors deepened. There was no pale lime or turquoise here. Tropic seas show these colors when the water is shallow over a light sand bottom, but here the surface water was deep jade, the color of the ancient greenstone Maori war clubs. And still the *pali* dropped away in the clear water until it finally faded out into the midnight blue of unknown depths.

I swam out of the bay and then turned left between the two towering rocks at the base of Hā'upu Peak. Directly beneath me, the sunlight converged in spokelike rays to a vibrating hub. Was I swimming over a hundred-foot depth or five-hundred? It was not a comforting thought. I drew my thoughts back to the surface and concentrated on the rhythm of the stroke. *One* two three, *four* five six. The pack floated easily behind, a reassuring tautness along the towline, fastened from the pack line to my waist.

I was stroking steadily, right hand hitting the water on one, left hand cutting in on four. Half my mind was on the scene below. The other half pushed back the insidious thought—sharks. I dared not admit the possibility. The snorkel made it unnecessary to turn my head to breathe, the fins gave added power. I lifted my face to check direction, then veered slightly out to sea to clear that next headland. The waves were getting higher and the chop more frequent.

I wanted to raise the masts and sail which, because of an onshore wind, I had left flat on the box until out of the bay. Now with the wind behind me, we could run before it, but I could not manage to rig the *Royal* while in the water. Rounding Lae o ka Pahu, the point of the drum, I looked for a place on the rock ledge to land, then made two stupid mistakes.

The first was that I'd forgotten to tie the mask onto my shoulder strap, the second was misjudging the surge around the point. As I grasped the lava ledge to climb up, a wave lifted pack and body, threw us roughly across the rock, then washed us back and off the ledge. Battered and gasping, I fought to the surface. The mask was gone, torn off my head and sunk. One fin had been wrenched off, but I found it floating. My head, arms and hands were gouged and bleeding. I didn't peel off the sleeveless shirt and jeans to look at the rest of me, but finned on into quieter water, climbed up to a smoother landing, then pulled the pack up and took stock.

It was nearly three more miles to Waikolu Valley; from there I could walk the rest of the way. There were a few places in between where I could walk along the shore, but there were no sand beaches and each shore landing in this rough sea would mean another beating. I decided to endure only the final one.

I jury rigged the shattered masts, raised the shirt sail, and set out. Without the mask and snorkel in the choppy water it was easier simply to hold my head up and use my arms to push the pack ahead, rather than to swim crawl stroke, tow the pack, and try to breathe on the side. It does not take much ingested salt water to nauseate me.

I headed out to sea, the shortest route, hoping to use the wind to push us, but the seas had risen to fifteen feet, so high that the sail did little good. We were in a trough more often than on a crest. I kept a steady flutter kick. The wind increased. It was going my way but the currents seemed erratic. From time to time the crest of a wave would break off and crash down on the *Royal* and me. I was still thinking in terms of square-riggers, and figured the wind at a force six on the Beaufort scale, about twenty-five knots.

Keep kicking, Aud.

There was a sudden slash of pain around my right arm. I looked down. The blue tentacles of a Portuguese man-of-war were wrapped in slimy cords from shoulder to forearm. I clung to the pack with my right hand, and with my left tore loose the long streamers and air bubble sac of the jelly fish. Wherever the stinging cells had touched arm or fingers were long lines of welts. They felt like the attack of a hundred hornets.

Keep kicking.

Was I even moving? I watched the crags. I passed the site of the cliff leap of two years before. Hadn't I learned anything? Well, you can't fall off the sea—unless one of these crests picks me up too high, that is. I bodysurfed between the rocks of Kūkaʻiwaʻa Point on the front slope of a lifting wave. On the left ahead was the two-hundred-foot rock column Huelo, capped with its windblown shock of *loulu*, the *Pritchardia* palm. I lined its peak up against the two-thousand-foot *pali* behind it. Slowly, steadily, I moved the peak along the head wall. I came even, then passed it. A mile to go.

An hour later at sundown I came around the last point and onto the rocks at Waikolu. It was a crash landing, but it was the last one. Kalaupapa was only a mile away now and all by land. I bathed in the river, then used the Hawaiian remedy on the stings by tearing my towel into strips, soaking them in urine, and wrapping them around the arm and hand. A poultice of papaya fruit would have been even better, but there were no trees here. There were seventeen other wounds gouged on the left hand. "If you're going to put yourself in idiot situations, idiot, don't complain."

There was an old pair of ragged pants on the bank. I peeled off the wet jeans that I'd been wearing for a little extra protection against rock landings and the water's chill, and put on the dry ones. Perhaps they had belonged to one of the patients at the leprosy colony, but the disease is almost noncontagious, and I was grateful for any warm dry clothing.

I built a tiny fire for *miso* soup, then soaked and applied the towel strips again. By now the medicine was developing a proper, more effective ammonia content. I crawled into a low rock and canvas shelter some fisherman had built. Propped on the one good elbow, I lay there and lit the last half inch of candle stub in a final wry attempt at the whimsy of sybaritic luxury. It was my fireplace, my candlelight-and-wine, but also my grim pleistocene defense against the sabertooth. I watched its flame lower, then gutter out as the wick slid down into the melted wax on the rock. I slid down too, and slept.

Next morning I shouldered the pack frame for the last time to walk the one remaining shoreline mile. My mind drifted. I smelled fresh figs.

"Sutherland, you have really flipped."

"But I smell figs."

"Sure, sure."

But there they were, on stunted windblown trees at the base of the sea cliff. Ripe, juicy, purple figs. Planted by a priest about 1900, they cling to life there in the wind and salt spray, and bear fruit each summer. I ate a dozen and carried more up the last steep trail.

I came through the planted forest of ironwood and eucalyptus to the grassed area of the park pavilion of Kalawao near the old village site where Father Damien lived and worked and died among the lepers eighty-five years ago. A group of picnickers was there, some of the Catholic sisters from the Kalaupapa hospital, and visitors from Honolulu, my first contact with people in a week. They invited me to lunch—cold juice, fried chicken, salad, ice cream. I accepted, and tried not to shove it all in at once with both hands.

Most of the visitors from Honolulu didn't know enough about the coastline to the east to be surprised at the trip, but one woman quavered,

"Weren't you afraid?"

I thought back—of cliffs and sharks and surge and boulders.

"Of what?" I asked.

"Oh, of being alone. Uh, some man might have come along, and, uh. . . ."

"If you mean, wasn't I afraid of being raped, no. If a man hikes five miles, swims ten, crash-lands on the rocks, and still has enough vitality to be interested in this grubby, salty body, he's welcome."

Sister Bridget chuckled. She had hiked over to Waikolu along the rocks many times, and had climbed down four hundred feet to the tiny lake at the bottom of Kalaupapa's extinct Kauhakō crater. I asked how she had managed in her long white habit. She laughed again.

"Oh, I just tucked it all up around my waist and snagged a pair of stockings into ruin."

Fine woman. They gave me a ride three miles across the peninsula to the small airstrip. Our flight out in the four-passenger plane went first to Maui, so we flew back along my seven-day route. It took three minutes.

That evening at home, I stepped out of the shower. Ann, washing her teeth, gasped. There were one or two inches of skin that were not black and blue and scabbed. Ten-year-old James stepped to the refrigerator, brought back a bacon label, and slapped it to my forehead. "Smoked, Sliced, and Cured in Hawaii."

The Canoe

. . . the wave lifted, curled with a hiss

It was three years before I went back. The beauty of the country still outweighed the hazards and was more important than the bruises. What could I learn to help me make an easier trip on Molokaʻi? What equipment could I devise?

Most of the experienced divers and boatmen I knew had seen so many sharks around the Islands that they could not share my lack of concern about *manō*. They urged me to use some kind of raft that I could climb on or into.

In fifteen years of skin and scuba diving in the Islands I'd only had one shark encounter, but the lack of such episodes was probably because I rarely used a spear. Certainly I was edible and very vulnerable, but without a string of bleeding fish or the unnatural movements nearby of a speared and dying one, I simply wasn't attractive to sharks. But there might always be that hungry or unpredictable one.

The clan and I often set lobster nets in front of the house —nets here in Hawaiʻi instead of the traps I'd used for commercial fishing in California. To haul the nets in and out through the surf and rocks, we used an inflated inner tube with a lashed-in center floor of plywood. Would it be more durable than the foam box, and could I climb up on it in case of emergency? I would still need the pack frame though, to carry the rig from Waikolu Valley to the Kalaupapa airstrip at the end of the trip. Also I'd need a pump to inflate it, or else carry the awkward thing all pumped up while hitchhiking from the Molokaʻi topside airport out to Hālawa at the start of the trip, looking like a giant walking doughnut.

I learned that there were alternate access routes into Wailau and Pelekunu, up and over the top from the south shore. There were no trails then, but some of the route was tagged with small nylon tapes tied to trees. The visibility there is fifty feet through the rain, and it usually rains. In many places the ridges are only two feet wide, but the *uluhe* fern might catch me as I slid off. I could take a machete and chop a route for ten hours a day. It has been done, by Lorin Gill, an Island mountaineer, by Hajime Matsuura of the Geological Survey, and a few other strong competent ones—far stronger than I. Behind a machete I don't have much weight and power, though I float and swim well enough.

Besides, if I walk into one valley with no swim gear, then I must perforce walk out, seeing only one valley and no coastline, and there's no way to walk into the sliced cleft of Pāpalaua, my mystic valley of rain and *moʻo* and the high waterfall. So it has to be a sea voyage.

Then in the spring of 1967, in the catalog of the Smilie Company of San Francisco, an outfit which specializes in mail order equipment for backpackers and mule packers, I found a listing and a picture of a small, French-made inflatable kayak. They called it a kayak, but it looked more like a canoe. The bow and stern were only partially covered and the width-to-length proportions were more like a Hawaiian canoe than a kayak. Besides, canoe or *waʻa* were terms familiar to the Islands. Kayak was part of a foreign culture. Would this tiny thing work for me? I studied the specifications.

It was six feet long with separate compartments of air on each side, tapering to pointed bow and stern, and was inflated by means of a lightweight, squashable rubber foot pump. An air mattress shaped the bottom and could be removed at night to sleep on. The double bladed wooden paddle separated into three parts for carrying. The whole rig, deflated, rolled into a bag twelve by eighteen inches and weighed fourteen pounds.

I ordered it and hoped it would arrive in time for the trip. I had carefully planned a leave from my job, and had scheduled my summer session university classes to allow for a week on Molokaʻi. I could do it that week or not at all that year.

As so often happens, time ran out before I had thoroughly tested the canoe. Chichester and Slocum had the same problem with their boats before going around the world, as did Byrd and Fletcher with their gear before Antarctica and Grand Canyon. Although the twenty-mile route along north Moloka'i was a small jaunt compared to their expeditions, I had problems enough.

Accustomed to inflatable objects which collapsed— tires, air mattresses, balloons, myself—I was wary of the new rig, so I brought along the foam box that had endured so well. It was the combination of the bulky box inside the tiny boat that was nearly disastrous on this third trip.

Bill Lacy, who lived up the road from me at home in Hale'iwa, and has flown into every remote area of the Islands in a variety of aircraft, offered to drop me off by helicopter at Puahaunui, the first peninsula, on his way to a job survey on the Big Island of Hawai'i.

How small the chopper looked at the airport in Honolulu; but I'd heard that Bill was one of the most competent pilots in the Islands, and I figured he was just as interested in staying alive as I was.

We flew low to the western point of Moloka'i, then lifted to nine hundred feet along the cliff edge and across the settlement of Kalaupapa. The cliffs were three times higher than we were flying now, and the rain blew in rivulets across the bubble shield. It was familiar country to me, yet new each time, and even more vivid so close in and at this height. I saw most of it through the camera's viewfinder, shooting the torrential waterfalls, the blue gray cliffs, and the dark misted Pāpalaua.

We landed at Puahaunui beside the walled open area of the *heiau*. Bill lifted off, and when the noise of his motor faded away in the distance there were only the sounds of the wind and the sea. I did not own a waterproof watch, so had brought none. It was sun and belly time from now on.

There was no joy that night. Was I tired or older or lonely, or was the transition simply too fast, from crowded

Honolulu to this isolation? A case of culture shock, I decided. But for the first time on these journeys, I wished for company.

The night wind and the rain came as always, but I was warm and dry in my lean-to, and almost comfortable now with the new air mattress from the boat. As I slept, one hand strayed out on the ground and unconsciously flicked in response to the hundred footfalls of the centipede that crawled across my finger. He sank his double claws into the tip; there was no mistaking that pincer bite and the pain, an occupational hazard to those who sleep on the ground in Hawai'i.

The pain subsided sooner than expected. I remembered the first time I'd been bitten. Then, a newcomer to Hawai'i, I'd stepped square on the center of an eight-inch centipede with my rubber *zori*, trying to mash him, but uncertain of which end did the damage. Was it the tail, like a scorpion? No. I learned instantly that it was the head as he twisted and sank his claws into my toe.

I was awake now and listening to the surf, apprehensive about the wind and rough seas. I wanted to read myself back to sleep, but my library had only three choices: a pamphlet on Hawaiian birds, a series of oceanographic articles, and a paperback of the lyric poetry of Edna St. Vincent Millay. It was too dark for bird watching, I was too wary to concentrate on science, and too emotionally vulnerable for Millay.

What I really needed was a super anthology for seagoing back-packers that would include philosophy, humor, travel, fishing, and hiker's gourmet cooking, with Hawaiian history, fish, animal and plant information. That would mean collaboration by Bertrand Russell, Farley Mowat, Sheila Burnford, Ballard Hadman, and Trixie Ichinose, plus Gavan Daws, Vernon Brock, Alan Ziegler, and Heather Fortner all in one book—a four-ounce waterproof paperback, dehydrated.

I fell asleep planning a dinner party for them all here at Puahaunui, along the lines of Hendrik Willem Van Loon's book, *Lives*, where he invites compatible people from the past and present. The invitations for Vernon Brock and Bertrand Russell would be carefully placed under that strange wooden deity over there under the *hala* tree . . . a potluck menu, freeze-dried. . . .

In the morning I found a quiet place to launch in the lee of Puahaunui peninsula. I pumped up the boat and tried to pack in the gear. The foam box would scarcely fit with me inside the boat, so I towed it along behind. I paddled around the ledge of Hāka'a'ano, then in toward shore at Pāpalaua, looking up toward the deep set waterfall of the *mo'o*. On the calmer side of the arch of Keanapuka, I stowed the paddle under the bow and stern and put on the fins and mask which had been lashed to the boat, then rolled over the side and into the water. I tied a line from my waist to the canoe and swam for shore, giggling under water. Did we look like a line of decoy ducks—me, the boat, and the box—all bobbing along? Or perhaps the humps of the Loch Ness sea serpent? I landed on the rocks with only a few bumps and reeled in the other parts of my monster. No wonder a shark wouldn't attack.

I yelled up the valley to whatever inanimate friends were still in residence, "Hey, hello. I'm home." Some of the driftwood lean-to boards of three years before were still there, carefully scattered in the jungle as I'd left them so as not to leave any obvious traces in the wilderness.

I set to work and cleared away the scratchy *hala* fronds, replacing them with fragrant ginger leaves, rebuilt the shelter, and bathed in the cold stream, clothes and all, to wash out the salt. I wrung out the clothes and hung them to dry by the fire.

Dinner was pork chops, fried pototoes, and applesauce, one of the pre-fab dinners now being packaged for backpackers; tasty, but infernally complicated by steps one through five of soaking this and adding that in numerous pots and cups.

After dinner I lay on the matted grass below the jungle, arms crossed behind my head, looking out to the north and watching the stars come out. The high black sides of the valley hid the view to east and west. I lay there thinking of the early Polynesian navigators and the contrast between my shore-hugging forays and their months of voyaging into the unknown.

Yet they were quite certain that there was land here. Birds came each year from the north, and left in the same direction. Land birds, not sea birds. Driftwood that had been shaped by man washed ashore, some of it with metal parts intact. Land and people were there in the north somewhere.

Polynesians had seen the Southern Cross revolving around a black void below the horizon. They had seen the Big Dipper swing around an opposite northern pit which they could aim for. There was no fixed Polaris to guide them in the sixth century. The earth's axis was tilted at a different angle to its orbit, and *Hōkū pa‘a*, our stuck star, then revolved around true north. But there were many ways to stay on course. At night they could keep the rising stars of Orion always at the same place on their starboard side. They knew the way back home by the series of known stars rising behind them, and knew how to compensate for winds and currents. They knew the zenith stars of each island. When Arcturus was directly overhead at its zenith, then they were in the latitude of Hawai‘i.

I sagged into the ground, my relaxed knee falling sideward, then roused myself and went back up to the shelter. By my rationed candlelight, I pulled the Millay book from the plastic bag. It fell open to "Exiled." She was "weary of words and people," "caught beneath great buildings," and wanted to "fear once again the rising freshet." Yes. That is the way I had been, but now I was here, wary as always, but ready for this place.

The candle was burning faster than last trip, a different brand. It would not last the week. "It will not last the night." Millay kept fitting in. Next time I must bring extra batteries for the new four-ounce Mallory flashlight. What a strange blend of her lyrical tender yearning and my abrupt pragmatism.

Next morning I went up to the head of the valley again, to the falls and the dark pool, but now the sun was glistening on the *ti* leaves and sparkling on the water. In the warmth of noon I shampooed, using the soap of old Hawai'i, the fragrant and gelatinous red ginger blossom, then rinsed off at the edge of the pool. Usually I swim under waterfalls, exhilarated, battling into the foam, trying to cling to the rocks behind the pounding white curtain, but I was still wary of this pool. The mythical *mo'o* were only sleeping, deep beneath the sunlit surface.

Back again in camp I decided to rig and push on to Wailau. The wind was fresh, but I could not hope for the relatively calm seas to continue. There were whitecaps but no big swells.

I tried to launch from the calmest place along the rocks, but with the pack in tow, even the three-foot shore break was too much. With just the pack it would be okay, or with just the boat. I tried lashing the pack on top of the bow, but it was top-heavy and appeared likely to capsize, so I paddled back toward shore, then swam and towed the boat in. We all took a beating on the rocks. I reached down in the water in the midst of the scramble, and with a quick finger twist, bent a perpendicular big toenail back to a flat parallel.

So I came back, set up camp again, and pondered all the possible methods of taking off. I was worried and I knew it. The next morning I allowed myself the use of real toilet paper instead of leaves, figuring that every small bit of morale building was going to be needed during the day. I decided to put all the gear in the boat, swim-tow it out through the breakers, then put the box overboard on a tow-line and climb into the boat myself.

As always, there were the many details to remember: put adhesive tape around the base of the thumbs to prevent paddling blisters; keep the gloves, camera, light meter, fruit bar, jacket, and knife accessible. I was thinking of wind shifts and the two thousand miles between here and Alaska. If I were blown out to sea, how long could I survive? How long would the boat last? I lashed the fins and mask to the boat.

We launched all right, and got the pack floating fine along behind, but it was a drag, a sea anchor. Two days before, coming down before the wind, it hadn't mattered, but here the wind was blowing toward the west shore of the bay, and the seas had a six-foot chop already. To get out of the bay and around Kikipua Point I had to paddle with the wind on my starboard bow. There was no keel or centerboard on the canoe to prevent sideways drift except for my own bottom curving down through the flexible hull. If I stopped paddling, we'd blow onto a lee shore, the terror of every yachtsman.

Paddle!

Over my shoulder I saw a sail approaching.

"Want a lift?" they shouted.

"Yes!"

"Try to get out to sea a bit farther."

As they tacked the catamaran around I paddled harder, then was scooped up by strong arms—boat, box, and all—onto the deck. They scarcely slackened their fifteen-knot speed. The maneuver was elementary to this gang, Carter and Emily Pyle of Lahaina, and two crew members of the winning yacht of that year's Trans Pacific Yacht Race. They didn't even say what the hell; paddling this treacher-ous coast in a six-foot boat was no more unreasonable than sailing it in a light racing catamaran.

We skimmed on down to Wailau, anchored, then swam ashore for lunch. While they sliced a freshly caught bonito into sashimi and opened a chilled bottle of German wine, I pried *'opihi*, the small limpet, off the rocks, and we shared potluck internationale.

Then they headed back the way they had come, up the coast for Maui, upwind and tacking, but with skill and manpower in a boat Carter had designed. Alone again, I launched next morning in the opposite direction with the foam box laid flat inside the boat. There was little space left; I was scrunched into the stern with my legs up on the gunwales around the box, a half-sitting, half-lying position hard on the belly muscles. I was figuring my systems as I went along. In all my reading about the sea, there wasn't any guideline for this kind of expedition. It was all trial and error, mostly error.

The boat was down by the stern, and there was no way to trim. It was raining, the dark clouds lying heavily on the sea. I stroked around the first point, Waiehu, with the wind becoming stronger and the seas steeper. Then the tops of the waves began breaking off. One lifted, curled with a hiss, and broke over my head, half filling the boat. I tried to bail and keep the waves in back of my right shoulder. I dared not look back at them after the first glance, when a twelve-foot sea, twice the length of my boat, came looming up overhead. I was half a mile out from shore, but there was no shore now, only three-thousand-foot cliffs dropping into the sea.

I paddled for two hours. Paddle and bail and paddle. It rained harder. When I was in the water, swimming, looking down at the rocks and coral below as I was lifted in the waves, then I wasn't afraid—I was part of it all. Boats were a different matter.

There was a mile to go to round the point into Pelekunu. A tern screamed down, its harsh cry shattering through the sounds of sea and rain and wind. Another wave broke over my head, nearly filling the boat, but it was still floating, responding to my paddle. I could not stop to bail and still steer to keep the waves in back of me. If we turned sideways, the boat would broach and roll over. With my weight forced back by the bulk of the box, the stern was almost awash. The next wave would swamp and capsize us. Turned over, could I cut the fins and mask clear of their lashing, put them on while holding the boat and pack, then tie the boat and pack together and lash the towline from me onto the boat? Perhaps I could. I didn't think so.

Tired, it is hard to get your breath in a strong choppy sea. A few mouthfuls of water and you gag. Exhausted, breathless, you quickly lose consciousness. I knew at last that I was paddling for my life, gasping and terrified. This was the worst.

Paddle! A whimpering, desperate grunt with each stroke.

The terror filled me to overflowing; it could intensify no further. Dig, stroke, steer, paddle for an eternal hour. Salt tears mixed with salt spray, but I could only keep paddling.

We came around the point, and the seas lessened. I shipped the paddle and bailed out twenty pounds of water, then started the long steady pull to the shore, difficult against the currents of the bay, but no longer dangerous. I paddled on, toward the tiny black sand beach in the sheltered corner. All else around the half-mile bay was rough lava and rolling rocks. I dragged the boat up the sand with the wave surge, then lifted the pack up onto the higher rocks. Paddle, wet gear bag, finsmaskandsnorkel, shoes— each item I lifted heavily up to higher ground, then the boat, then I lay on the rocks, limp and trembling. After half an hour I could move again.

I put some rocks on top of the boat to keep it from blowing away, and stowed the paddle and fins inside. I tied the rest of the gear to the pack frame, and started across the boulders toward the trail on the far side which led to the shack. Along the shore I broke up driftwood and carried it in my hands for a cooking fire for supper.

Coming to a curve on the trail, I stopped short. The hairy black pig stopped too, ten feet away and waist high. I yelled, half in fright, half to frighten him away, and clutched my jagged sticks tightly. He came toward me a few paces, uncertain, weaving his head to get my scent, his tusks curving upward. I grabbed for the camera, but he had turned and run back through the brush. "Camera first, idiot. Yell later!"

The trail was washed out under the *hala* leaves and between the roots. One hole dropped a hundred feet to the rocks below. I had mental images of the pack catching on the roots while my legs waved frantically below.

The shack was in even worse shape than three years before. With tools and equipment I could do a lot toward repairing it. I put a layer of rusting metal scraps on the rotten floor, four rocks to hold my grate, and made a tiny fire to boil a cup of water for soup, then sipped it while hunkered in a dry corner, my back against the wall. At least I had a wall to put my back against. I wasn't assailed from all sides.

It was getting dark and the heavy rain came through the roof in thirty places. One bunk was a little less wet than the others. I put my plastic sheet on the top bunk to catch the rain, laid my boat's air mattress on the lower bunk, and in damp jacket, damp jeans, and dry socks, lay down to sleep, not quite warm enough and not at all dry.

It rained all night, hard, then eased, then came pounding again, hammering against the walls and pouring through the roof. When I slept I had intensely clear, dramatic dreams. When I woke every hour or so I was terrified. The sea roared, and then the rain drummed louder than the sea, and things went bump and screech in the dark.

Finally the sky lightened. I threw my plastic sheet and some boards up onto the roof, climbed up the cliff behind the shack, and hoisted over to the eaves. With the boards to hold the sheet down, I made one end of the shack somewhat rainproof.

A cloudburst with lightning and thunder came tearing across the bay in a dark wall. The lightning was nine seconds away, six seconds, three. I tensed, waiting for the blasting shaft, but it passed on. I could not see the beach, and three torrents between here and the shore made the trail impassable. Maninini, the pouring-off place, is what the Hawaiians called this cliff to my right. Had the boat washed away?

Of one thing I was certain. I would not leave here for a while. I would not go on my own unless the seas became very calm. Even the ignominy of being rescued was preferable to another four miles like yesterday's.

I tied my pink bra upside down to the top of a bamboo pole stuck in the ledge outside above the catwalk. It was the only bright color in all my gear and perhaps would serve as an international distress signal. Should a boat come close enough to see what it was, they might be curious to investigate further. Below it was lashed my jacket. Both flew in the wind, but the only planes were high and far away, and no boats passed at all. A wry grimace. "You got yourself here, Aud. Now get out."

I rested and wrung out my clothes in the one dry end of the shack. How could I best prepare for the night? The sun was out only five minutes all morning and even then it rained steadily. The whole bay was brown with mud from the streams. I rationed my food, not knowing how long I'd be here, but was not really hungry.

The one creative satisfaction all day was making a stove candle. I found a jug of kerosene in the sagging toolshed and put some in a jar, then punched a slot in the lid and inserted a folded wick of old gunnysack. It worked fine, shed a bit more light than a candle, and slowly brought water to a boil.

At sunset, after four hours of lighter rains, the bay was almost blue again. I squished along the path to the shore and brought back the boat. The rain had filled it to the brim, and the weight of the water and the rocks had kept it from washing away. For dinner I ate an excellent Lipton beef stroganoff. My morale improved.

I repacked, figuring how to arrange the gear for the morning departure. What could be left behind to lighten

78

the load? A small voice stood off and said, "But promise, Aud, that you'll try it only if the seas and wind are reasonably calm. You know you have no heart or courage for this."

A wave hit the ledge below and threw spray up thirty feet to the level of the shack. I held a pair of green nylon bikini underpants on a stick over the kerosene wick to dry them out to wear under the jeans as another layer for warmth that night.

In years to come what would I remember of all this? The terror, or the splendor of the mountain across the bay this evening? At sunset the top of the mountain reflected a shining bronze from its wet leaves and rocks, while the shadow of the ridge and peak behind me made a silhouette of gray gold halfway up. A rainbow arched over it all, brilliant against the darkening clouds, and out beyond a mile of heaving sea a second, lower, rainbow outlined the black rock of Mōkoholā Island.

I was not sure that I would go, but I woke before daylight and went outside. The wind had been blowing hard all night; half-awake I had heard it screaming and felt the shack quiver, but now it was almost quiet. There were stars between the clouds. I hurried to be ready by dawn. By nine the wind would rise and the seas would be choppy again.

By the light of my homemade smoky lamp, I carried out my own terse orders of preparation. Leave a note to greet the next user of the shack, or to someone who looks for me in case I don't make it today. Climb up to the roof to retrieve the plastic sheet to use for shelter if needed that night. Take only two meals. Leave the rest so as to lighten the load.

Lower the pack by a nylon line from the catwalk in front of the shack. Drop down the unbreakables. Carefully let down my very breakable self. Inflate the boat, and shove the packsack in under the bow. Turn the foam box on its side, so as to push it farther forward than before, making more room for me and a better trimmed balance.

Lash everything in tightly so that if the boat capsizes I can just right it and tow it along, even full of water.

Daylight now and gold in the east. Wear jeans and a shirt for warmth in the water and for a little protection when I land on the rocks. Put on the socks to pad the ankles, and then the fins. Don't tie the fins and mask to the boat; have them on, ready to swim. Spit in the mask and smear the saliva around, then rinse it in a tide pool so that it won't fog up on my face from my breath. Check the fastening of snorkel to mask. Put the mask on my forehead and tie a line from its buckle to my shoulder strap. Put bandages on my thumbs, pull on the gloves; my hands are not calloused enough to withstand the constant friction of the paddle. Lash a line from the paddle to the boat, and another from the boat around my waist. Should we capsize, the wind could carry an untethered boat away faster than I can swim after it. I'm learning.

Now. Pull the mask down over eyes and nose and clamp the snorkel in my teeth. Hold the boat and all on one thigh. Step down from the ledge to a jutting rock, awash a foot below. Ten feet to the right the surge crashes up into a crevasse. Surge and backwash. Surge and smash and spray twenty feet high. I threw a piece of wood in here last night and watched it float back and forth. It was not sucked in, but I have had bad experiences with ledges and am wary.

Look to the left. The waves swell across the bay and wash around my knees. Wait for the lull. What lull? I drop the boat into the water and jump into it. The boat tips and I roll out the far side. It's no time to be funny, Aud. I grab the stern, and power kicking with the fins, push it out away from the ledge. Then floating up to parallel the surface, I grasp the far and near gunwales in each hand and slide across and in. Pull the double paddle out from beside the foam box. One detail forgotten. The pull apart paddle is together, but the blades are neither parallel nor at right angles. The joints are cold and corroded. I cannot twist them. The boat rolls and pitches. The waves crash into the crevasse. I push one blade forward and wedge it with my

80

fin against the pack and twist the other blade. It creaks into place.

Now paddle for the first point. The sun is up.

Paddle!

It is only a quarter of a mile. Halfway there already. I do not look back. From somewhere comes a sudden mosquitolike high whine. Air is leaking from the boat. I reach under the hull to the valve but all is in place. The side does not seem to be losing much air. Just paddle. Planes fly with one engine gone. If it loses all the air on one side I can tow it on half a boat and the inflated bottom if necessary.

Round the first point, Pau'eono, literally, the death of six. Don't make it *ehiku*, seven. The two rock stacks thrust upward. I want photos but didn't loop the camera around my neck this time, nothing extra to hang and tangle. Go between the rocks. The one on the right rises sheer a hundred feet above the white froth. Steer around behind it, hoping the lee will cut the chop. The wind builds. Paddle for the next point, Pahu, the drum, another half mile. I wedge my outstretched legs up and over the round sides trying to keep out the slop. The green swim fins stick upright on the gunwales.

Round Pahu Point into quieter water. Pass the ledge of last trip's disastrous attempt at landing where I lost my mask. Keep paddling. Head in toward the waterfall of Hā'upu Bay. The water near shore is murky from the heavy rains. It is not a sand beach by the waterfall as it looked from a distance, but only a boulder shore. I land easily, find the tiny leaking hole, open the pack, and apply a drop of fast-drying glue. Get out the camera and take delayed-action shots of a grinning voyageur, sooty from the kerosene lamp last night and this morning. The most dreaded part is past.

I head out to sea again, past Kapailoa, the long lift, and look into the lava tube tunnel. I've heard of boats motoring from one end to the other, and I would like to try paddling through, but the water is boiling around the entrance, and the rise and fall seem to suck into the gaping mouth.

At the next point, Kūkaʻiwaʻa, there is a thirty-foot gap between the rocks. I surf through it on the leading slope of a breaker. Below my left paddle, the lifting wave sucks out, baring the gleaming yellow seaweed on the rocks. Past the two-hundred-foot Huelo—black, phallic. Head for the last point, Leina o Papio, a mile to go. Tiring now.

I think I can—the children's story of the little engine. Count one hundred strokes. Count again. Every fifth stroke is a correction to bring the boat back on course. Sometimes the chop is so high that held momentarily on a crest, I cannot reach the water on either side.

Hoe aku i ka waʻa. Paddle ahead the canoe. Lift the paddle, reach, dig, pull, and finally you'll get home. A tricky stunt won't do it, a sudden burst of speed won't help—just lift, reach, dig, pull. Each word is grunted with a stroke of the paddle.

Hoe – aku – i ka – waʻa. Over and over.

The blisters break. It does not matter. *Hoe! Hoe!* I come around the last point to Waikolu and start yelling. *Hoe.* Yaaaay! *Hoe.* Yaaaay! A shout of triumph all the way to the shore. I land, haul up the gear, laugh, yell. I made it. I made it. By myself.

I gather *ʻopihi* to take to friends, then carry the forty pounds of pack a mile along shore and get a ride with some of the kind people of Kalaupapa across the peninsula. The small plane arrives and we fly to Honolulu International Airport. A telephone call brings my daughter. Waiting for her, I drink a quart of milk straight down. Cars honk, tour buses blast out their black exhaust, the canned voice reiterates, "This is a three-minute passenger loading and unloading zone, please do not leave your car. This is a three-minute . . . ," but we head out toward the North Shore and soon leave Honolulu behind. Through pineapple fields and sugarcane acres and small towns I come home again to the long old house by a quiet sea.

Return
to
Pelekunu

. . . paddled toward the peak of Haʻupu

I had to go back again. To be that terrified of anything, that incompetent, survive by that small a margin—I'd better analyze, practice, then return and do it right.

I was at work now, sitting at the desk in my skirt and stockings and heels. The wounds were healing and the bruises fading. What did I learn? The gear would have to be tested and tried in all combinations before any more trips. A boat was not yet my element the way the water was. I had missed seeing the life of the underwater world. Fear was more a barrier than the problem feared.

Before 1962 I hadn't known enough to be frightened of this coast. In five years the three trips had taught me how difficult it could be, but I still wanted to go back. It wasn't because of the "challenge." I didn't feel daring and I didn't think my character needed to be improved by conquering something, but now I knew the magnificence of the place, strong and fulfilling. Was there some masochistic satisfaction in the beatings I had taken? I thought not. It was simply that the tender power of Moloka'i was a far more vivid and compelling memory than the physical pain.

At home, asleep at night on the wide bed by the fireplace, I would waken from a nightmare of being jobless and alone in some mainland city, trying desperately to get back to Hawai'i and the children. Reassured by the moonlit wooden beams overhead, the deep authority of the winter surf sounds outside, and the softly burning coals of the fire, I would think of Moloka'i, how it would be in that light and with that surf. I would pad down the hall. The children breathed quietly and safely in their rooms. Some new idea of rigging or gear or timing for the next voyage would come to mind before I drifted back to sleep.

I made short swimming or boating expeditions near home, or with a son or daughter on other islands, trying out equipment. I learned to trust the little canoe, and, as a substitute for the foam box, I found sturdy canvas and rubber waterproof bags which I could stuff and lash under the covered bow and stern of the boat. Even full of cameras and food and pots they trapped enough air to float should they be torn loose. Just to make sure of the buoyancy, or in case of a rip in the outside bag, I put a chunk of foam into each one along with the gear which was prepacked into double plastic smaller sacks.

I studied the catalogs of equipment put out for hikers, made my own lists, narrowed them down, and planned ahead. I had been a first aid instructor for many years. I reviewed it all again, and taught an advanced course, the best way to really learn.

One thing I never quite managed was to be in top physical condition. Each time I paddled out from Hālawa I swore, "Next time, more push-ups," but there never seemed to be enough time for jogging and swimming. More likely I was just too lazy and not quite vain enough to work continuously on turning flab into muscle, but the 66-inch, 120-pound body was a good responsive tool, and I could put it to hard use.

My job as an education administrator for the army was demanding. The complex conferences, telephone calls, speeches, rapport in individual counseling—plus all the demands of a family—seemed to require brief occasions of time alone for balance. Moloka'i was the complete contrast. Whether I had just come back, was about to go, or was in the middle of winter with the big surf in front of the house a reminder that such an expedition was impossible then, I longed for it.

So I studied the maps and the photos, talked to people who had been in the valleys, and did research at the libraries and archives. Each summer I chose a week's vacation from the job, and when I planned for Moloka'i, it was Pelekunu I was going to. The tools, the building materials, the books that I took were all for living in the

shack there. Piece by piece I bought waterproof equipment
—camera, light meter, flashlight, watch, binoculars, mas-
cara, glue.

The preliminaries were the same each year. Apply for
leave time, make sure the job was covered, get someone to
stand by for the clan in case of emergency. Assemble the
gear, double-check the lists, pack it all to carry on the
plane. Take the earliest flight to Moloka'i, contact the arranged ride or hitchhike out to the east end. Repack, inflate the canoe, launch out through the surf and paddle.

Did the Hawaiians who lived in these valleys for centuries have such rough seas under their outrigger canoes?
Worse. They were here all year 'round, while I am only a
summer tourist. And they made their canoes and paddles
and food sacks by hand, without metal tools, while I only
adapt the products of technology.

They were here as early as thirteen hundred years ago.
In 1969 a team from the University of Hawaii and the
Bishop Museum, led by Tom Riley, Patrick Kirch, and
Gilbert Hendren, excavated in Hālawa Valley a fully
established community with numerous house sites, ter-
races, and *heiau.*

Using radiocarbon dating and a relatively new tech-
nique of obsidian dating based on crystallization rates in
volcanic glass, they dated the settlement at approximately
A.D. 570, the earliest occupation period yet known in
Hawai'i.

But for Pelekunu there is no known record of when the
Hawaiians first settled. Perhaps the answer lies buried in
some shelter cave or within the stone walls of an old house
foundation. As pottery is used in other parts of the world,
layer by dirt-covered layer, so fishhooks with their varying
patterns of shapes and fastenings have been used as chro-
nological indicators in Hawaiian archeology.

My excitement and anticipation rose as I headed west.
Past the mists of Pāpalaua, across the wide expanse of
Wailau, under the wet crags of Waiehu, I paddled toward

the peak of Hā'upu that marks the west side of Pelekunu. I was a whale homing to its summer island, a sea turtle migrating to its own small bay. The sun sparkled through the rain and unfurled a small, personal-size rainbow low over my canoe.

In years to come, when I recall the sounds and scents that hold the essence of these voyages, among them will be the harsh, poignant "skark" of the noddy tern. Suddenly there will again be sun and rain, whitecapped blue sea and white surf on the black cliffs, the sweeping trade winds and the rocking boat—and out of the sky the soundfall of the tern's cry. On all the previous trips, I've looked up, waved a salute, and called out, "Hello, old friend. *Aloha kakahiaka*, I'm back." Ananoio, the cave of the tern, is the name for the last long sheer scarp from Waiehu to the eastern arm of Kaholaiki Bay.

Strange how the geographical themes appear and recur. The eastern thumb spur of Wailau, called Kahawaiiki— little river valley—is a small version of its larger parent valley. The small bay of Kaholaiki is a perfect U the same shape but one half the size of the next bay, Pelekunu.

Four rock islands guard the approach to the two bays. The outer ones at first glance are ship hulls cleaving bow waves through the seas. The two inner ones, Mokumanu and a smaller rounded one are sea stacks, hard cores of rock remaining after the softer surrounding lava had eroded away. They are the Scylla and Charybdis of my departures from Pelekunu. Coming in, I see them there, sentinels waiting at the gates for my exit, but I look on into the bay to the ledge on the west side to see if the old shack is still standing after the heavy winter surf.

Above the shack is Hā'upu Peak. Many versions exist of one persistent story which tells of Chief Kape'ekauila, a descendant of the earliest migrants from the Marquesas. About A.D. 1200, he is said to have established his kingdom on Hā'upu and surrounded his fortress on the peak with natural cliff barriers and hand built rock walls. The legend tells of stone-paved paths leading westward over the ridge to Hā'upu Bay. Kape'ekauila and his forces in their red-stained outrigger canoes ranged widely, and finally, on the island of Hawai'i, abducted the most beautiful woman in the Islands from her husband and children. The chief held her for eighteen years in his fortress until he was slain by invading rescuers led by her sons.

The Helen of Troy plot is long lasting. So too is the name for the west side landing ledge on old maps and new, my launching place below the shack—Wa'a'ula, the red canoe.

I paddle on toward the eastern corner of the bay, wondering if there is a chain lying in the depths of Pelekunu Bay like the one I found back in the jungle at Wailau.

Probably the first Europeans in Hawai'i were those of Captain James Cook's expedition, who "discovered" the Islands in 1778. For two hundred years Spanish galleons had been sailing from Manila in the Philippines back and forth to the west coast of Mexico, but there is as yet only slight, tantalizing evidence of their having seen or stopped in Hawai'i. Is there a Spanish anchor below me, encrusted with coral and half-buried in the black sand and rocks twenty fathoms down?

Cook himself only sailed past Moloka'i, noting it in his journal, but not landing. Other early explorers' references to Moloka'i indicate that it was a neutral area of secondary importance, used as a political issue among the chiefs of the larger islands.

The wars between the islands, the increasing influence of foreigners, the beginnings of commerce, all of these bypassed Moloka'i. Contact with the Westerners after 1778 was minimal because of the lack of a deep draft harbor on the southern, sheltered shore. Communication with the northern valleys was even more difficult. Only the missionaries, both Catholic and Protestant, felt duty-bound to contact every isolated group of native Hawaiians.

In 1832, Harvey Hitchcock, a Protestant missionary

stationed at Kalua'aha on the south coast, wrote of a population of twenty-seven hundred in the thirty miles of north Moloka'i coastline. Fifteen years later another missionary, C. B. Andrews, wrote from the reports of his assistant, Mr. Gulick, of a hundred and fifty inhabitants in Pelekunu. In those fifteen years most of the north coast people had gone. Some had been lured away by the activities of the new towns on other islands. Many had died, victims of the foreign diseases to which Hawaiians had no immunity. Mr. Andrews estimated that from 1846 to 1848, one out of six in Hālawa Valley alone died in an epidemic of whooping cough.

The Protestant missions continued their religious efforts and also started schools along the north coast. In 1847, one teacher was at Wailau, one at Pelekunu, and three at Kalaupapa. But the ministers lamented the gradual decline of their church attendance. For the residents the novelty had worn off. The missionaries could seldom get into the valleys to preach because of rough seas and rainstorms. They could not stay during the winter because it meant leaving their families alone at the south coast station, and they couldn't bring their wives along because of the "fearsome seas and difficult landings which no woman could endure." Nor could I have, with an ankle-length skirt and ten petticoats.

There were too few ministers for too big an area; Catholic and Protestant missionaries were competing for the few souls who remained. As I come to the corner of the bay just outside the line of the surf, and sit in my canoe waiting for a lull in the wave sequence, I think of another paddler who did the same. In the summer of 1875, Father Damien, priest, carpenter, caretaker of souls and bodies at the leprosy colony of Kalaupapa, built a chapel in Pelekunu, apparently with lumber from an almost new chapel at Kalaupapa which he demolished. Damien's superiors in Honolulu were critical of his actions, taken without direct permission, but the fathers in Honolulu surely had no concept of which size lumber would fit into a native canoe, the difficulties of landing it through the breaking surf of Pelekunu, and the necessity of getting it there during the two summer months before the winter surf rose.

My journals laconically record a variety of my own landings:

9 July 1968: The usually calm east corner had a steep crosswise shore break. I came over to the center, waited for a lull and paddled hard. Stepped out on black sand at low tide and hauled up the boat.

4 July 1969: Came ashore in the east corner at 11:00 A.M. No problem. The calmest water of any trip so far. Now after I'm finally learning to manage even if the seas are rough, they've been comparatively smooth. Too bad I couldn't have exchanged today's flat seas for those of that first terrifying boat trip in 1967.

. . . watching the clouds and the view across the bay

As I come ashore
I always think, "Two days, five, a week here out of a whole
year. How short it is, for these are the days I'm doing what
I most want to do." I haul the boat up onto the rocks and
begin dividing the load. I cannot carry it all in one trip,
since I've brought material to repair the shack and to enjoy it, not just to survive.

I straighten up from the pack, look around and listen to
my valley. The sounds are those of the rounded boulders as
the waves roll them up the slope of the shore and back
down again—like a ten-lane bowling alley. No matter
where I am through the year, I can recall that high-pitched
click, clunk, clunk of the rolling stones and the shoosh of
the foaming water, and I am back at Pelekunu with all its
space and solitude.

The sight of a fairy tern—high, white, and soaring
against a cerulean sky—makes my throat catch. I smile
ruefully. There is no way to describe the ethereal fairy tern
in lean prose.

I look up the valley and see the ridge between here and
Wailau. Perhaps next year I can stay long enough to climb
up to it, but even from there, what could I see? Only trees
and jungle, no trail, a few tags on the trees which might
mark the route, but also might have been placed by a pig
hunter following his dogs, leading nowhere. It would be a
fine three-day expedition to hike over the top, down
Wailau, and swim back here.

All of this part of Moloka'i, say geologists Gordon Mac-
donald and Agatin Abbott in their book *Volcanoes In The
Sea*, was formed by a volcano, one of three that made the
island. The first one poured out a series of lava flows a
million and a half years ago to form West Moloka'i. As it
ceased flowing, a larger eastern volcano kept erupting until the two merged to form a single island. A small third
one much later made the peninsula of Kalaupapa. Up

there on the ridge is the overgrown eroded caldera of East Moloka'i.

My equipment is sorted now and the first pack load is ready. I half-squat, lift the pack frame to the top of one thigh, slide the right arm under the shoulder strap, then flip one hip toward the pack and stand with a single swiveling motion, tossing it up and onto my back. I slide my left arm under the strap and cinch the hip belt tight. The rest of the supplies can wait for a while here in the tangle of beach morning glory and rocks, well above the surf.

Walk cautiously. I haven't yet cut my usual guava walking stick for tripod support. Across the shore rocks, easier than the tangled vines, to the base of the angled path. Breathe deeply, place the feet carefully on the steep slope that drops to the sea; the trail is muddy. Slide around and scramble all you want when there's someone to yell to for rescue. Not here.

I pause and check the light meter for a reading, then set the exposure on the Nikonos and am ready for any tusked pigs, but only goats appear around the curving trail this time. I take two pictures before the black female gives a bleating honk of warning to the kid. She sounds like a seasick diesel truck.

Along the trail I renew acquaintance with each remembered tree and bush. Here's the 'ākia, whose dark bark was stripped for cordage by ancient Hawaiians, the roots and leaves crushed to stun fish in the river pools. There aren't enough 'ākia trees left nearby for me to try it.

I pass the ridge which leads up the west side of the valley and thence back to the high plateau in the center of the island. It would be the route out, were I not committed to a sea voyage by the weight and bulk of the boat, bags, pump, fins, and paddle.

The trail winds over a fallen log, into a gulch with cool shade and green ferns, then out around another ridge. Ahead is Hā'upu Peak, and down to my right is the shack, its roof a jumble of torn plastic sheeting, loose boards, and the scraps of the original roofing paper. That roof is so critical that I always inspect it first. When it lets rain pour in, the floor, the bunks, and everything else inside disintegrate with damp rot during the year while I am gone.

Now the trail comes to the stream which is the water supply for the shack. It has cut deeply into its own small gorge and I must leap across, pack and all. An uprooted sisal plant has slid across the trail beyond the stream, so I climb up and around it, sidestepping the stiff dagger spines, then follow the switchback down across the grass to the shack. The door is open and hanging by only one hinge. I look in and groan at the mess, slide the pack off and start cleaning.

The equipment that I left last year, partly to lighten the load in the boat, and partly in preparation for my return this summer—the food, utensils, bedding, carefully packed in plastic bags and lidded buckets—is scattered or missing or left out to rot in the rain. Someone has left a pot of oatmeal thick with mold, a pile of stinking fish heads and guts, and part of a goat carcass full of maggots. I burn and bury and heave into the sea, apologizing to Kanaloa, then fashion a broom from *hala* leaves and sweep out the goat and rat droppings, thinking in the same terms about the last visitors.

I unpack some of the gear, repair the broken hinge, hook the door behind me and sit on the steps, making a project list of things to do in the next few days. I lean back, grubby and tired, and smile at my world. Alone again. The freedom of solitude.

Despite all standard advice, it does not seem foolish to come alone. I now know how and what to do, having learned most of it the hard way. The few competent, compatible people that I'd like to have along are seldom available to come when I do. Alone, I am doubly careful. At home the family knows my planned route, the deadline for my return, and what to do if I don't meet it. I carry signal flares for emergency; a daily sight-seeing plane flies by a mile away.

There is another reason to come alone, besides my own need. The children, by necessity, have been trained to self-sufficiency. We have no television—they read omnivorously. There have been few other children living nearby, but there is plenty of life in the tide pools in front of the house. I work days and often nights, and have the car with me; the kids ride bikes or walk or run. There are no organized playgrounds; they've learned to skin-dive and to surf, Jock becoming "number one" in the world. We have the list posted of twenty things every kid ought to be able to do by age sixteen, which includes fix a meal, splice a cord (manila or electric), change a tire, change a baby, listen to an adult with empathy, see work to be done and do it—that last one will take about five years more.

But now the clan is growing up; the youngest is a teenager. I won't know whether I've done a good job raising them until they all reach forty or so. Then I'll see how they respond to other people and to their own children. The transition is going on. They need the opportunity now and then to be without me, to make their own decisions, to make their own mistakes and repair them.

I need transition too. There is a strong feeling of "what next?" In June 1968 I finished paying for the house, saw the third child graduate from high school, and completed a master's degree—an eight-year part-time project. My wise and concerned mother set up a fund for each child to help with their college costs. Three would be in the university in the fall with part-time jobs. Three up and out and one to go. I've been doing vocational counseling as part of my job for many years. Will my own job be one to keep me happy when there is no longer such a need for as high an income as possible? Will I like living more simply and alone?

One thing for sure. The world is changing, the changes gathering momentum and spinning out in all directions

98

like a giant fireworks pinwheel. What next, indeed? Josephine Tey, an English mystery writer, once wrote the thoughts of a gentle English lawyer with his routine tea and biscuits, "Is this all there is going to be?" So now I am living with four distinct individuals, no longer children, whom I like as people, despite their being relatives. It is high time I stop using motherhood as a prime reason for existence, as a basis for my own value, and go on to something else. I know that I do not want to live alone always. Children of all kinds are dear and important, and there are many ways I can be useful.

It is time to stop being afraid, time for nightmares to cease. Am I not learning to do Moloka'i better? Can I not learn to lose other fears? Can I even learn mathematics? I have been fighting change, digging in with both heels, dragged along, yelling protests about a changing world. I laugh aloud at the image, then go inside and fix supper.

At dusk I sit content inside the cabin. It is the shack until I have cleaned it and lighted the candle; then it becomes the cabin. A moment ago I stepped outside and walked across the wet grass to the edge of the wall next to the catwalk. The sea and rocks surrounded my feet, the peaks and glittering sky were an infinite universe above. Then I turned and saw the candle through the window, the light in the wilderness, the reassurance that there was one small refuge in it all, enough for one small human.

I come in and close the doors to keep out the wind, but I can see through the window the top of the peak across the bay.

I continue the journal, terse penciled scribbling, written while sitting at the rough little table, without the lapse of time and space between here and home to soften the impact or empurple the prose.

9 July 1968: Left Wailau this morning in far calmer water than last year. Even sunshine. No foam box this time, just the bags stuffed and lashed in the boat. Must have

left the air pump on shore at Pāpalaua. Couldn't find it, so topped off the sides and bottom of the boat with a few lungs full. The waterfalls are all small and Wailele is dr. . .

A sudden gust of rain hits the roof. I leap up and peel bare except for the black tennis shoes, tuck the plastic sheet under my arm and the flashlight between my teeth, and climb up to the roof by way of the toeholds on the cliff behind the cabin. I lay out the plastic and weight it down with boards and rocks. It will not survive a hard wind, but for now it keeps the bunk end of the cabin dry, and I can sleep without waiting for the drips on my face to force me up there in the middle of the night. I come back down, dry off, and dress again. It's much easier to dry me than the clothes; that's why the quick strip. It would have made an unusual flash photo.

Carpentry becomes a bit more organized in 1969. By now it's an annual trip. Here I am again.

I am gloriously lazy all the first morning leaving the shack messy with my gear while having a deluxe breakfast of oatmeal with dates and brown sugar, orange juice, and coffee with whipped "cream." Then I belly under the shack amid the broken glass and rusty cans to see what I can find. There are some good tongue and groove one-by-six boards, but I also find that the floor joists of one sagging end of the shack are lying directly on the wet ground and rotting away. I don't think I can lift them alone, even by levering them up with an old length of pipe, as there is not enough space under there to pry a lever. I'll need a house jack.

I go to the beach to look for more driftwood planks for the roof, lash them to the pack frame and return. Hmmm, I'd forgotten. Tongue and groove is ⊏ ⊐, but there is also shiplap ⌐ ⌐. I have some of each that don't mesh; they are of varied lengths and half-rotten, but the work progresses even so. I brought along a folding Sven saw, but

100

most of all I need two or three rolls of the old familiar ninety-pound tar and gravel roofing paper that I've applied at home so often. No way can I bring that in the six-foot canoe.

At two o'clock a boat goes by, a mile at sea. Through the binoculars I make out the name *Aukaka*, Dave Nottage's cruiser. Good fishing, Dave, my upside-down distress signal isn't hoisted today.

By late afternoon the roof is finished, as nearly watertight as I can make it with the materials at hand. At least it has fairly solid boards, and the plastic sheets that I've tacked down will last a little while.

Grubby and sweaty, I get out the signaling mirror and a comb. Signal mirror, hah! That's pure pretense, Aud, and you know it. All the directions I've read for signaling with a mirror are totally complicated. Besides, like using a magnifying glass to start a fire, when you need either one is during an emergency when there probably isn't any sun. Okay, the mirror is just for vanity, so that I can create some semblance of female back at the edge of civilization. But right now, egad! The natural look is for the twenty-year-olds. All this nature only gives me soot, squint lines, and puffy eyelids. Pluck that wayward eyebrow with pliers and go scrub.

The trail to the stream is not one to skip along. Four inches wide, sloping outward, and slippery with wet grass, it snakes precariously along a steep slope with a rock ledge thirty feet below.

Beside the pool I stand and soap, then rinse off from a pot of scooped water so as not to suds up my friends, the two dozen catfishlike 'o'opu that share my *furo*-size tub. They look at me with swiveling eyes and fan their pectoral fins. As I slide into the cool water it is startling to feel them on the bottom. More startling to them. I stretch out, floating, head resting on one mossy edge and toes touching the other end, watching the clouds and the view across the bay, then climb out and stand exultant and tingling, and laugh out loud. This is the aim of the upper income residents of Honolulu—a vacation home on another island, with tropical landscaping, a sunken tub, a view of the sea, and no telephone.

I dry on an old diaper that serves as towel, pillow, pot holder, scarf, and bandage, then pull on the jeans. I go out on the catwalk, hoping there will be a break in the threatening clouds and a clear sunset view.

The first structure on this landing was built by Johnny Wilson, later to become engineer for the Honolulu Pali road and then mayor of the city. He and his wife Jenny, a former court hula dancer, lived in Pelekunu from 1900 to 1914. She was postmistress here, and he made the village his headquarters between contracting jobs elsewhere in the Islands.

In 1900, there were about seventy people in the valley, including forty children in the school which had been taken over by the newly annexed territorial government. Wilson built a derrick with a sixty-foot swinging crane on this Wa'a'ula ledge below my shack to handle people, taro, and supplies coming in and out of the valley, since the waves breaking on shore were just as much problem for them as for me.

In 1915, Jack London, his wife Charmian, and young Dr. Kenneth P. Emory, present dean of Polynesian anthropology, were swung ashore one at a time in a basket hung from the crane, while the others waited on the deck of the boat rising and falling in the seas below.

By 1919 most of the villagers had left the valley. A few old Chinese taro farmers remained, who worked the flat wet fields near shore with water buffalo from dawn to dusk, and smoked opium during the quiet evenings. From time to time a small sampan from Kalaupapa would chug in to the base of the derrick at Wa'a'ula ledge and wallow there in the swells, loading the bundles of taro corms for the Hawaiians at Kalaupapa to pound into poi.

During the 1920s, dealers in Honolulu underbid the

Pelekunu growers and shipped taro into the settlement by interisland steamer. Finally the derrick was dismantled, the buffalo were turned loose, and the old farmers bundled up their few belongings to board the small sampan, looking back at the quiet valley with its golden peaks that they would never see again.

Some of the village houses were taken apart and shipped out. Big surf and tsunamis cleared out the remainder, but the buffalo multiplied and kept the trails open up the valley to the Geological Survey station until hunters came in and tracked them down for meat.

The G.S. teams needed a shelter closer to the shore and a way to land when the ocean was rough. In the 1930s A. E. Minvielle, Jr., a Honolulu engineer and surveyor, brought a crew and built a cabin at Waʻaʻula ledge and a new firmly anchored steel catwalk cantilevered out over the water with a rope ladder hanging from the end. These too were abandoned when the compilation of data on water resources of the valley was considered complete.

Pelekunu is deserted now. Once every two months the G.S. teams go in by helicopter to the high upper valley to gather additional rainfall information. An occasional hunter or fisherman lands on the shore, but after I leave weeks go by with no one there.

Pelekunu Plaza

. . put on a clean white T-shirt for dinner

In 1970

I make plans for a longer stay and for an all-by-sea voyage in a new nine-foot inflatable canoe. I am like a sailor who suddenly gets a forty-five-foot ketch to replace his thirty-foot sloop. Such space and luxury. I am becoming the commodore of a Tupperware navy.

My new yacht weighs eighteen pounds, so I plan to carry it on my back as little as possible. On the other hand, *it* will carry several hundred pounds, so I pack tools, books, extra clothing, two bottles of wine, a one-pound down-filled sleeping bag, a separate air mattress, a hammock, a small hydraulic axle jack, a big bath towel, two hundred feet of garden hose to pipe water from the stream, a faucet, and heavy clear plastic to make window panes. Everything *and* a small kitchen sink.

I paddle my loaded craft out from Hālawa, wondering where the Plimsoll lines are on a nine-foot boat, and remembering all the other shacks I've hauled gear for. As a kid I built tree houses in the mountains each summer, standing by as the cabin was opened and the storm shutters were unbolted, then lugging them out into the forest to hoist up as platforms into yellow pine or Kellogg oak. Do kids do it any more—build an aerie in the branches and haul up a bucket of books, crackers, and peanut butter, so they can read and munch and look down through the leaves on the world below from a secret hideout? The family cat and I both went wild that first day of summer—free of the city . . . and *all those trees to climb.* When my cousin Sid came to visit we made Indian tepees, pirate caves, knights' castles, and corrals for our broomstick broncos.

The carpentry continued when my own children were small and my husband was out at sea. The old beach houses at Surfside and Hale'iwa needed repairs and remodeling, and I often woke up in the morning with a pencil still over my ear. I am as yet only a cobbler, a wood butcher, but I've slowly learned about miter boxes, ratchet screwdrivers, and not to open paint cans with a wood chisel. There was a long history of nest building before Molokaʻi.

After leaving Hālawa I paddle all the way to Pelekunu with no stops along the way. The new boat handles well even with the load and in choppy seas, but the journal also records:

> 21 July 1970: A rough bad landing. I was overconfident. The boat half filled with a breaking wave and nearly capsized. But as barnstorming pilots say, any landing you can walk away from is a good one. Curious goats watched in amazement.

I divide the gear into three loads for the pack frame. The first view of the shack from the trail above evokes a gasp of dismay. In December there had been a convergence of storms from the North Pacific, and some of the highest surf ever recorded hit the North Shore of Oʻahu where I live. It also hit Pelekunu.

The outer half of the catwalk has been torn away. The cables which supported it snapped back and tore off half of my roofing job of last summer. Now the boards and the heavy plastic sheets and the cables hang in shredded scraps and swing in the wind. I do makeshift repairs that evening and am asleep on my feet as I stagger to the bunk, hammer still in my hand and flashlight in my teeth.

In the morning I make a quick switch from sailor-carpenter to hydraulic engineer. I tear out the remains of the rusted sink and install the new one, bolting it to the old sink board, and draining it with a cut length of bicycle inner tube through a hole in the wall and into a trench down over the cliff. I put one end of the garden hose into the highest pool of the stream, weight it with a rock, and screen the end. Then I lay it along the slope, down to the cabin, up and over the kitchen windowsill, clamp it in place, and screw on the faucet.

Then, turn it on—wait—HO! The water comes trickling out, then flowing free. I make a couple of gleeful capering leaps—and fall through the rotten floor.

For three days I work on the floor, tearing out two sagging bunks and all the old boards of the floor, then use the axle jack to raise the floor joists. I inch along on my back, shoving big rocks for foundations in place with my feet, lower the joists onto them, and make new flooring out of the old bunk lumber and driftwood. When I need a carpenter's level, I have the longest, bluest one in the world. I need only get down on my knees, butt up, and squint across my boards to the horizon.

In the toolshed are hanging the remains of the old harnesses that the G.S. crews left from the days when they used mules to haul equipment to the upper valley station. I'll leave them too. Those scraps of leather and metal belong here.

I install new windowpanes by cutting the clear plastic into squares and tacking the edges with staples onto the old frames. I patch the roof again and again, build shelves and repair the bunks. The old weathered boards are a pleasure to work with—paint long gone, warmth and history slowly achieved. I know people who are like that.

Each time building material is required it is strangely provided. I have no drill, but when I need a hole for a bolt to shore up a wall, there is a knothole not seen before. I need a four-by-four floor beam, twelve feet four inches long. Sure, Aud, and would you like a twelve-foot man also? But I look out in back of the cabin, and leaning against the cliff is a four-by-four, twelve feet two inches long, that I make do. I do not question my unseen elves, my *Menehune* helpers, but only look up, wave, and say thank you, *mahalo*.

Winter storms have torn away the wonderful old outhouse until only the beams spanning the chasm remain, but on shore, tossed high on the rocks, the seas have brought a replacement seat, the shattered side of a boat with two perfectly placed portholes. It is a project for another year. The sea and the land provide.

So too, do they provide food. As the repairs are completed I spend more time foraging and preparing creative additions to the menu. There is fruit in the valley, guava, banana, pomelo, *hala*, and java plum, but they are not all ripe at once, and you have to know where to look and how to prepare them. There are coconuts, taro, fish, goats, wild pigs, and a variety of seaweeds, *limu* in Hawaiian.

'Opihi with escargot sauce is a standard menu item now. I bring 35-mm film cans from home stuffed with the mixture of butter, garlic, and parsley, then pry two or three dozen of these small limpets off the rocks. They aren't easy to find in Pelekunu. Neither the rolling rounded rocks of the shore nor the rough lava, *'a'ā*, are a good habitat for them. After scrubbing off the bits of rock, I lay them shell down on a grill, slather a blob of sauce on each one, and hold them over hot coals just until the meat comes free and the sauce bubbles. I squat by the fire, lift the grill off and let the *'opihi* cool enough to pick up with my fingers, then slide each tender morsel into my mouth and lick the shell clean. I wipe my fingers on my jeans; it gives them a buttery waterproofing.

For one person, however, foraging is a full-time job, so I do it only as a supplement to my own supplies. Quite possibly I could use up more calories searching for food, hauling it and preparing it, than I'd get back eating it.

There are still some prejudices to overcome. I ought to be able to eat anything that anyone in the world finds edible, but some of the accidental experiments aren't very successful. Cockroaches have a greasy bland flavor and a crunch that is still repulsive. Ants are crisp and acid flavored, and cling to my tongue even after being chewed.

There are more wild bananas in Wailau Valley than here in Pelekunu but there is now a banana stalk planted by the back door, and some day I may have them—simmered in butter and honey and flamed with rum. Guava jam maybe, coconut pancakes. . . .

I don't hunt at all, although it would be better for the valley flora if most of the goats were eradicated. I am grateful to the various private land owners just to be here, and I don't know their feelings about the goat population.

Thanks also to others who care for the cabin as I do. Last year I arrived and found my note from the year before in a jar as I had left it, "A.S. left for Waikolu 10 July 1968." The place was neat and clean, a prop was placed under the roof, and the doors were carefully latched to keep out the goats. Added to my note was another, lettered with charcoal. "JB and AB left for Oahu July 22, 1968." Others must have been there during the year and left it as clean as they found it. When I arrive this year I find that some one has left yards of blue plastic sheeting which I promptly use for the roof.

Guests are coming more frequently to the Pelekunu Plaza in the last two years. It is time to leave a permanent guest register—in a waterproof bag, of course. All of those who arrive and leave under their own power, by hiking or swimming or paddling, seem to appreciate Pelekunu and the cabin, contribute to its upkeep, and leave it clean. Half of those who come by motor power are welcome guests also. But no one has come while I am here. I have only my own thoughts and the animals for company—a real variety of animal life.

It is about 3:00 A.M. I wake from a dream and hear the seas rising, but something else awakened me. There is a bug in my ear. He crawls across the eardrum, his footfalls sounding and feeling like a branch scraping on a tin roof. I roll off the narrow air mattress onto the bare boards of the bottom bunk and fumble for the flashlight. The bug's antennae are probing. I grope through the plastic bag of miscellany for the bottle of olive oil, tilt my head, pour a teaspoon of oil into the ear, slosh my head around. After two doses he stops squirming, and I tilt oil and bug out onto a towel. It is a small, greasy, expiring cockroach. Olive oil is very versatile. I use it to fry fish, clean my face, dress salads, treat sunburn, lubricate zippers—and drown bugs.

I tumble back to bed and to sleep, and an hour later wake to a rustling sound over by the sink. I flick on the flashlight, directly into the red eyes of a huge rat. Well, rats aren't anything new, so I bang my shoe on the floor and frighten him away from the food supplies, then bang again when I hear him scratching about in the dark later.

By daylight the rat is gone, but there is a different sound at the kitchen windowsill. I turn over to look and a mongoose leaps out the window. Mongooses were brought to the Islands from Jamaica years ago by sugar planters to kill rats, but it didn't really work. Both are predators, but rats are mostly nocturnal, mongooses diurnal, and they only seem to say "Hi" in passing as one goes home to sleep while the other heads out for the day's hunt.

Two hours later there is a wild squealing, and the mongoose, with a live rat in his mouth, runs out from the back of the cabin and down over the cliff to a pile of rubble. So I'm wrong, and perhaps the importation was justified, but what else does the mongoose eat, what does the rat keep in check, and what checks the mongoose?

As the construction work nears completion, I make an up-valley hike to look for some of the plant and animal life that was here before any of the humans, Caucasian or Hawaiian.

The old G.S. trail has disappeared, overgrown with the *uluhe* fern, bamboo, and ginger. Only a here-and-there goat path or a muddy pig freeway remains. I crawl under the ferns or flop on top of them, get up and walk on the mashed fern, fall prone again, and proceed one body length at a time. Finally I retreat to the river and walk along beside or in it. The rare sun is warm and the stream is cool and shining, with pools for swimming, small rapids that glitter and splash, and high falls dropping from the side cliffs to shower under.

In the upper valley, three miles from the shore, the rains are more frequent and the flora is quite different. More of the old Hawaiian plants remain; the imported guava, sisal, mango, and java plum are not seen up here. *'Ēkaha*, a four-foot-wide clump of bird's nest fern, nestles in the crotch of a *kukui* tree. Adder's-tongue fern, *pololei*, and lichens cling to the limbs of the *'ōhi'a* tree between the puffs of red blossoms. Symbiotic relationships between plant and plant, plant and animal, are well developed.

There is no four-language nomenclature for the endemics as there is for most of Hawai'i's biological items. *'Ōhi'a* is indigenous, and though it also occurs in Tahiti and New Zealand, it is known only by its Polynesian name and the scientific one, *Metrosideros collina*. It is the predominant tree in most of the rain forests of Hawai'i; ferns are the most common plants of the lower growth. The *'ōhi'a*-fern symbiosis is mutually beneficial. The *'ōhi'a* has a shallow root system; it needs the ferns as ground cover to hold the moisture in the soil, and to keep the soil itself from washing away in heavy rains. The understory ferns like the shade of the *'ōhi'a* and the mulch of its dropping leaves.

'Ōhi'a is a tough hard wood; the genus name *Metrosideros* is from a Greek word meaning heart of iron. So resistant is it to rot and insects, that it is used throughout the Islands for fence posts, even in the rainy ranch country of the Big Island of Hawai'i. At Ogden, Utah, in 1869, the golden spike that connected the first railroad across the continent was driven into *'ōhi'a* ties.

"The *ohia*-fern ecosystem is the most important habitat of the majority of the surviving endemic forest birds," says Andrew J. Berger, who describes and deplores the loss of the extinct, the rare, and the endangered species in his book, *Hawaiian Birdlife*.

The Drepanididae family of birds are also native to Hawai'i, and occur nowhere else. During the centuries when they had no enemies, they increased to a point of keen competition for the available food supply, and made gradual physical changes in their own bodies and beaks to gain access to nectar, seeds, grubs, and insects. Endemism, evolution, and the adaptation of species have been demonstrated far more clearly in Hawai'i than in the Galapagos Islands, but Charles Darwin never visited Hawai'i.

Some of the rare birds may possibly remain in the backs of these valleys, and I always look for them, squatting there with the mud squishing in my shoes and the rain dripping down my neck, plastic bagged binoculars and ears swinging like radar saucers to catch the movements and sounds.

I look for other things too, the old village sites, the terraced walls of the taro patches of the ancient Hawaiians. And if I find something, a burial cave, a stone adze, a fish god, as I've found in other parts of Hawai'i, I shall do as before. Cover it and leave it where it belongs.

As I come back down the river, an 'auku'u, the black-crowned night heron, is my escort. It is a young adult with a soft brown plumage instead of the sharp black and gray of the mature bird. Each year I've seen one here—seven birds, or the same bird seven times. But no, it would have adult plumage by now. It flies ahead, stops until I catch up, then springs away and flies another hundred yards. When we reach the shore, it circles, "whawks" down at me, and flies back up the valley.

I shade my eyes and look after it. How long have they been here and where did they come from? Any bird in Hawai'i before man had to fly at least 2,000 miles or else come drifting in on a bit of flotsam, like me. Three years ago in the sandy ledges of western Moloka'i, Joan Aidem, a Moloka'i resident with an untiring scientific curiosity, found the 25,000-year-old skeleton of a flightless goose. That find poses more questions than it answers. How long was it here before the flightless characteristics evolved? What other remains are as yet unfound, of both bird and man?

I wade into the river to cross it again, stopping to do a fast laundry job on the muddy clothes and body. I take off the shirt and bra, unzip the pants and roll them down to my shoe tops, soap everything, then lie in the current and roll around on the rocks to rinse off mud, soap, and leaves. I stand and pull up the jeans, put on a wrung-out shirt and bra, and climb up out of the river, then cross the shore rocks to the smaller stream at the west corner of the bay. Angled back a hundred yards into a thicket of coconut, banana, croton, and bamboo, is the village site of former populations, but there are few remains. The tall shafts of the coconut palms blow in the wind above the other trees; below them the nuts drop and sprout anew in piles of fallen fronds.

Eating coconuts is no simple matter. They don't drop off in the neat hard balls that cartoon monkeys throw around. I didn't even recognize as a coconut the very first one I ever saw in its husk, washed ashore on the beach in California. The husks are thick, and although my son James can jam one onto a sunken pickax and strip it down to the nut in ten seconds, it takes me half an hour. But I have no pickax, and unfortunately the best nuts for drinking are the green ones that remain attached high overhead until they dry. The old Polynesian cultures had seven names for coconut, according to the stage of development and use. I need a pickax, machete, and grater, plus strength and skill to climb the trees.

A vine is woven into a strange piece of curving metalwork. It is the end support of a Domestic treadle sewing machine. Over there rusting into the leaves is the broken corner of an oven door from a wood burning stove. Very recent artifacts left behind when the houses were moved out or abandoned in the 1920s.

I stub my toe on something hard. It's a metal stake,

sharpened to an edge, firmly and deeply set into the ground for only one purpose—husking coconuts. Not a golden spike, but someone up there likes me for sure—four-by-fours, knotholes, portholes, and a husking stake. It's becoming delightfully eerie.

But I have a date with a mountain, and so I trek back to the cabin and ceremoniously set the table for a formal dinner with a bouquet of white ginger from the valley and fern place mat. Before this trip, I finally bought an Optimus backpacker's stove which burns white gas. Once primed and warmed it announces its efficiency with a comforting small roar. I doubt if I'll ever go back to makeshift kerosene cans, or cooking on a smoky wood fire in the wind and rain, wearing my diving mask and snorkel to keep the smoke from smarting my eyes and nose. Tonight I wear the mask for another reason. I'm chopping up the last fresh onion in my supplies, and this way it's tearless.

I put the pot of stew on to simmer, pour a glass of wine and go out to the nylon net hammock, strung from a corner of the cabin to the *hala* tree. Straddling carefully, I ease my shoulders back and my feet in, then go limp, totally contoured and supported, weightless like floating on a cloud.

It is time for the "shadow of the mountain," the golden time of the day when the lowering sun burnishes the earth —green gold on the slopes, molten blue for the sea, and gilt black on the rocks. The shadow of the Hā'upu Peak behind me falls first on the swelling surface of the bay, then slowly moves across the water and up the whole mountain on the far side. A rainbow arches at the upper right, repeating the shape of ridge and shadow. Night does not fall; it rises from the earth. The shadow reaches almost the top of the ridge, then the sun is gone, the shadow gone, and the sea already dark, but the clouds are still pink and white and the sky still blue.

By candlelight I bathe in the pool, then put on a clean white T-shirt for dinner, a charming size-44 frock which

116

comes halfway to my knees. The stew and wine are served formally—at a decorated table, with knife and fork instead of fingers.

Year by year and day by day I'm getting over the fear remembered from the first three trips. I am becoming very brave in the daytime when the conscious mind is fully awake and functioning. It is only when I half awaken in the dark and am not armed for combat, when my seams are unglued and hinges undone, that I get the 3:00 A.M. willies. How different this place is when I am relaxed and have plenty of time and the weather is good.

The second morning in 1972 starts with rain pattering on the plastic roof, plinking off the eaves onto a tin can below, and slanting against the black rocks behind the cabin. I stay in bed, outsleeping the rain and reading. Then the solid overcast separates into clouds moving west, the peaks emerge, and the wind begins to whip the sea out by the islands into whitecaps. The colors intensify and the mountain across the bay seems higher, a dazzling world. It is a day for sailors and not a boat has passed in six days.

I stand outside with a cup of coffee and look down to the ledge and the chasm where I have so fearfully launched when leaving Pelekunu. Surely I can learn to swim there without fear. The water is clear now and the surge no worse than around the Waimea Bay rocks near home where I skin-dive so often.

Grimly I take finsmaskandsnorkel and climb down to the edge, put them on, and jump in. Ho, it is lovely. I can see down about thirty feet to a dark rock, then follow it out of sight as the water deepens. I hang and drift with the surface surge, then flip one foot up behind me, bend at the hips and drop down to peer under a ledge, then turn and twist through a tunnel and back up toward the surface. From beneath, it is an opaque blue glaze, a chiffon scarf undulating softly in the wind, separating the two worlds.

Spouting through the snorkel, I lift my face and see the

117

black rocks and the cabin, and I am lonely, a mortal. I turn back under the sea, suck in my middle and wash across the shallows of a ledge, then have a moment's vertigo as the underwater cliff drops off to the darker depth. No longer alone. *Kala*, the horned unicorn fish, is there; so too is *hīnālea*, the vivid orange and green checkered wrasse. I am not swimming in the sea. I am the sea. Deep in a crevice, I hold to the rocks waiting for my eyes to become accustomed to the darkness, then see the shadowy waving antennae of a three-pound lobster. I have no glove, it is out of season, and I doubt my ability to get him anyway. Besides, if I am the sea, I cannot eat my friends who dwell with me.

Back at the ledge under the catwalk, I hold to a lava knob, strip off the fins, and leap with a surging wave up onto the shelf. I climb up to the cabin and lie on the rock wall. The trade winds chill my back and the rough rocks warm my belly. This is joy.

What next? Across the bay is another place I have feared. Coming down into the sea from the high ridge are the little falls and streams I struggled past in previous years, trying only to reach shore. I have learned that the boat can take much rougher waves than ever I thought it could. My journal records the beginning of this year's trip:

> 4 July 1972: Left Hālawa at 5:30 A.M. It becomes the worst weather I've known, seas fifteen feet, wind thirty knots. Tops blown off the waves and foam scudding along the surface. My elbows chafe on the sides of the boat and my side aches from digging in the right paddle to hold us on course. The heavily laden boat lifts to the crest of a wave and flexes down at both ends as if to break in half, then slides down into the trough and rises again.

David Lewis, who sailed alone in icy seas and hurricanes from Australia to Antarctica to South Africa, and was dismasted twice, has formulated a fine tongue-in-cheek principle. Lewis' Law states that for every point the wind increases, the boat diminishes one foot in size. At that rate I was paddling a two-foot boat, but at least my seas were warm. And I was less than a mile offshore, while he was at the place in the world that is farthest from any land.

I try for a tiny vestige of pride by formulating a corollary to Lewis' Law. Aud's adage: Wave size is proportional to boat size. A twelve-foot sea to my six-foot canoe would be like a sixty-foot wave to a thirty-foot boat.

By the time I stopped for lunch at Wailau valley the seas had calmed. There I met Mrs. James Naki, daughter-in-law of the Wailau walker of my uncle's days, her nephew, and five grandchildren who were all there for a week of fishing. Mrs. Naki was born in Pelekunu and raised in Wailau, a beautiful, barefoot, tough old woman, truly a grand lady. The boys, Kenneth, Manny Boy, George, and two others, took my boat out. They were instantly competent in handling it, and jumped overboard in waist-deep water to carry it on their shoulders up onto the rocks.

But now the seas are far flatter than the Hālawa beginning, so I inflate the boat and paddle across the bay to explore the waterfalls that drop to the sea on that calmer side of the bay, sheltered from wind and wave by the high ridge. I lash a towline around my waist, dive out, and climb with the surge up to a ledge. As I pull the boat in I stumble and fall backward, and just lie there laughing and use my flattened body as a soft "boat ways," hand-over-handing the canoe up my length to a higher level. Giant 'ape grows in the stream there above the waterfall, each leaf an "elephant ear" four feet long.

I paddle on around into Kaholaiki Bay and gather taro leaves, lū'au, from the base of the head wall, then swim out of the bay over the eroded pockets and caves tunneled through the basalt, towing the boat behind me. I fin along, out past the enfolding arms of the sphinx that I had looked down upon from the plane so long ago.

Arriving at the cabin this year, I landed for the first time with the canoe on the ledge under the catwalk instead of paddling to shore, deflating the boat, and packing it along the trail. It was reasonably successful, and so I try again for practice. Timing is essential. I sit in the boat, mask over my face, just outside the surge, watching the swells over my shoulder, digging in with the paddle blade like a pectoral fin to keep in place. Now. Paddle four strokes ahead, fast and hard. Slide the paddle under the bow, roll out and swim over to the ledge, watching the cliff rise and fall under water. Wait for the rise, grab a knob, turn the finned foot sideways to place a heel, up and out. The wave drops away. Wait for the next one to lift the boat, grab its side with one hand and the tangled towline with the other, back off and up to a higher ledge. Okay. With no load aboard, it works well.

I go to the stream to wash off the salt. There is another hazard. The stream is mossy and steep from pool to pool. I had always stepped carefully and braced against a fall. Hmmm. I sit at the edge of the top pool, shove off, and carom neatly down the slide, landing with a giant splash in the tub. The 'o'opu leap frantically, then settle back into the ooze.

A new idea. I go back for a bucket, then scoop and slosh all the leaves, ooze, and fish down to the lower pool. I scrub the moss from the sides and bottom, rinse it well, and let it fill with clean water. I have no qualms about disrupting the 'o'opu. Unlike salmon, they migrate to the sea to spawn, then return to their freshwater streams. With their suckerlike belly fins they can climb the four-hundred-foot cliff of 'Akaka Falls on the Big Island; so tomorrow they will no doubt be back on the bottom here.

My pool is clear and sparkling, but how shall I heat it? Add Japanese-style *furo* to that list of projects that starts with rebuilding the outhouse. Oh yes, and add a fireplace for the cabin. Aud, you're becoming a damn developer!

By candlelight I cook dinner, eat and read, then finally blow out the candle and sit in the doorway, sipping

Benedictine and watching the night. Across the bay the beam from the Kalaupapa lighthouse, six miles west, flashes from right to left at ten-second intervals, and casts onto the mountain there, shadows of Hāʻupu Peak and Mokumanu Rock different from those of sunset.

Living alone here at Pelekunu is such a strange blend of animal wariness, simplicity, beauty, and idiotic glee. Two elements are ever present. Always it has just rained, is pouring down, or the dark clouds are forming over there beyond the ridge. Always the seas pound against the cliffs. I learn to judge the height of the waves from the intensity of the roaring sound.

What is it like here in the winter when the giant surf comes up? A mile from my house in Haleʻiwa I once saw the waves "close out" at Waimea Bay, when they reached thirty-five feet in height and broke solid all the way across from one headland to the other. How would that be here? I think I would retreat from the cabin and watch from farther up the mountain.

There should have been an eternal photographer here, with a time-lapse camera like those that take one frame every ten seconds, and thus speed up the opening of a flower into a ballet of continuous motion. He would be suspended in space here and take only one frame each year, to record the first underwater eruption, the first tip of the island emerging above the sea, the intermittent eruptions and flows, with the flaming molten lava dimming to gray rock. Then the erosion beginning, the plants appearing, the fish and birds arriving. No, one frame a year is too fast. With one shot every ten years it will still take three hours to show my movie of the two-million-year history of Molokaʻi.

In years to come someone else may sit here and muse and watch the night, and feel the unseen presence of an ancient Hawaiian in his *malo* of tapa cloth, a white-bearded Chinese with his opium pipe, and a tousled blonde in tennis shoes. My daughter Noël wrote a note one evening when she was ten years old and home alone, saying she was "plain old happy." Me too, here.

A full moon is rising over the mountain's left shoulder. What have I been doing, cowering in the corner bunk each night with all of Van Gogh's *Starry Night* out here? I bring the air mattress and the down bag out onto the grass. So it was not just the glory that the thin, bearded painter saw, but the vast wheeling across the sky, the movement through the night of the stars and planets spinning on their axes, whirling through their orbits. I watch and sleep, until the rain on my face sends me back inside. But as I step through the door, I look back across the bay. The setting moon has created an unearthly miracle. Arched above the mountain is a moonbow—a lunar rainbow—of pale colors etched on a dark cloud.

Then the rain drops a curtain across the bay.

How rare and lovely to wake up only when my mind and body are ready. No sudden sound awakens me, no traffic or alarm clock or people, only the end of sleep. The cool sea air comes in the open window, the birds sing, the waves slosh across the ledges and the clouds float over the peaks. Light rain drips off the eaves. The down sleeping bag is so light and warm—like sleeping in a cheese soufflé. I smile and stretch, and finish reading the last book, Charles Seib's, *The Woods*.

I'm a reader, not a writer; a looker, not the creative artist, but how I do appreciate those good ones. There are books that belong here, so that their creators can be here vicariously to enjoy this place. Do writers ever feel the places that their books are? Do the words transmit the understanding and the enjoyment of the reader back to the author? What books should be here in the cupboard, what pictures on the wall? I shall bring Hayakawa, Robert Capon, Tom Neale, and John Graves with me next time.

Now on the last day, I go up to Hā'upu ridge, climbing the steep crumbling basalt, remembering the mountaineer's admonition. "If you fall off a mountain—other than a sheer drop, that is—roll onto your back. You can dig in with your heels better than with your fingers; your back is more protection for your vital parts than your front, and sliding on your back you can see the ledges or clumps of brush to steer for—sort of." I have no intention of trying the theory. I'm pure coward on Hawai'i's wet and rotten rock, and I plan to stay that way.

I climb first to the ridge directly above the cabin, straddle it, and sit braced by my right foot, with the left foot dangling over an 800-foot drop. Looking east I can see the waterfall of Wailele, the wall beyond Wailau, and the peninsula of the *mo'o* Kikipua.

The paved paths of Chief Kape'ekauila are nowhere to be seen, though there are rock walls up near the peak. Climbing through the guava brush on the ridge, I stop in midstep, startled by the gaze of a goat. She lies there on her side, slender hooves tucked under, ready to leap if necessary. I give her time to move away, but she stays, silent and serene. I step close enough to touch, speaking softly, glancing at her and then out to sea. Sally Carrighar, author-naturalist, says we should not stare at wild animals; as with humans it is an invasion of their privacy. The pupils of her eyes are not cat-vertical nor human-round, but horizontal thin ellipses across the large, lemon yellow circles. After long moments she gets to her feet and moves slowly away.

At the top of the ridge I look west and see all the places that I shall know so intimately at sea level tomorrow. "Kapailoa, Kūka'iwa'a, Huelo, Ōkala, Kalawao." I chant their names aloud, an incantation to invoke good weather and quiet seas.

Paddle
On

. . . out of the blue, into the yellow sunshine

If I get up
in the dark I can be ready to go by daylight. I want to be
well along the route by nine when the wind picks up and
the seas become choppy. The plane leaves Kalaupapa at
two. Everything I don't need in the morning is packed
now. I blow out the candle and slide into bed, then set the
mental alarm clock. Select a circuit, program it with the
image of a clock face and the hour hand moving slowly
around from nine to five. . . .

In the darkness a switch clicks. I flash the light onto the
small waterproof wristwatch. Five o'clock. I lie there
reluctant, then pull the plug on the air mattress, unzip the
bag, and swing out. While a pot of water heats, I pack the
sleeping bag in with the gear that must remain dry, one
camera, the film, the plane ticket. After coffee and *miso*
soup, I pack the cup in the bowl, the bowl in the pot, and
the pot in its own soot bag, then distribute the weight between the large bags.

I take everything out onto the grass, then look around
the cabin one last time. Will it still be here when I return? I
latch the kitchen door from the inside, go out the front
door, turn the wooden toggle, and pat the door a gentle
farewell. The cabin itself is mine now, a gift from the
Geological Survey people, and also by right of carpentry,
but shelter there is freely given to all who take care of it. I
would like to own the land too, but even if the legal rights
were mine, I would not possess it. But it owns me, I belong
there. I am beholden to Pelekunu.

I lower the equipment. You ought to know how to prepare and launch by now, Aud. This is the eighth trip. The
clouds turn pink, then gold.

I flip the loaded boat into the surge, leap in, and paddle
away. Threading through the slot between Mokumanu
and Pau'eono, I pause and photograph the black obelisks

against the mauve of sky and water. Of all pictures possible along this coast, I would like most to have one of the tiny boat floating here at sunset, with the shadow of the mountain and the rainbow, huge and golden there beyond the bay, but there is no way that I can run and leap, paddling, into that self-portrait in seven seconds.

Around Lae o ka Pahu, I paddle into the quieter water, looking for a sea cave that I missed previously. There it is, an old lava tube where the molten rock drained out a million years ago leaving the hardened shell. I paddle two hundred feet into the mountain and sit back there in the dark womb, holding with my fingers to the sleek rocks at my shoulders, hearing only the soft reverberating boom of the waves, looking out to the tiny patch of light at the entry.

I untie the underwater flashlight, pull down the mask, and roll out of the boat, probing the depths with the single beam, but see no octopus or lobster. The sides of the tunnel are smooth, no crevices or caves to shelter cephalopod or crustacean. Always in dark water and at night, there is the certainty of what the beam illuminates, lighting the front ten degrees of the circle around you, and the uncertainty of what lurks behind your shoulders in the other three hundred fifty degrees.

I tow the canoe out of the sea cave, put the flashlight back, and swim on through clear green water, watching the scene below, until I become chilled, so climb into the boat and zip up the jacket.

Then *hoe aku*, paddle on, on to Kapailoa, the long lift, the sea tunnel through the cliff that I've looked at before but never had calm enough seas, or courage enough, to paddle through. The tide is low now, the swells rise and fall, but there is no boiling chop. I peer in but can see no exit; so stroke around to the other end. Ah, it makes a right-angle turn in there, and the ceiling seems high enough all the way.

Okay, intrepid, let's go.

I paddle back around to the entrance, shoot into the maw with the terns screaming a protest overhead and the surge bouncing the boat from one side of the tunnel to the other. I come to the angled turn. The color, the blue grotto color, the brilliant, opaque, acetylene torch blue color reflecting up though the water from the sunlight at the entrance! Brace the paddle between my fins, read the light meter, set the aperture of the Nikonos camera, and take pictures as a wave surges through the entrance, lifts the boat to the ceiling and drops it back. Push off the rock sides with my hands, change the camera settings, and shoot again, body and mind drenched, saturated, *pulu pē* with water color.

I paddle out the exit, out of the blue, into the yellow sunshine and go on. Approaching the peninsula of Kūka-'iwa'a, I move in close to the rocks and look up, remembering the first journey in 1962 and the leap from the cliff. What have I learned since then?

Confidence mostly, some techniques, the use of suitable equipment. One day I may try to write a book on how to do it, but it might be too specialized—swimming and inflatable boats in warm seas. Paddling fiberglass kayaks in the wild rivers of Idaho as Montana naturalist John Craighead does, purposely upside down half the time, wearing a face mask to see the trout and other fish, sounds like a different set of skills. Betty Carey of Queen Charlotte Islands paddled cedar dugout canoes in British Columbia, but they have icy water and seven-foot tides.

I had to keep trying until I learned how. Maybe I could learn almost anything by being an absolute idiot the first 999 times. Could I learn to adapt a sail for this boat? The bamboo masts of the *Royal One* might puncture my inflatable. How about the ultimate in folding, furling sails— an umbrella? I nearly capsize giggling at the mental image of whipping down this coast with half of my double paddle tucked under my right elbow for steering, and my left hand and toes bracing a flowered parasol. Okay, idiot.

There are many alternatives I've found. No one system is the ultimate answer. If one route is blocked off there is another way to go. I've learned to live without things and alone. The ability to live in a variety of styles, city or country, with people or without, in different languages and cultures, with enthusiasm for the small luxuries, gives me a power over the future, whatever chaos the world comes to.

There is a sensuous joy in being alone—delight in the simple animal pleasure of blowing my nose with one knuckle, peeing in the moonlight, and trying a Tahitian dance step with only myself to snicker. There is a smug ironic satisfaction in finding an ingenious solution to a problem which was caused by my own inadequacy or stupidity.

Men and women are more alike than different. Women too need to feel the coyote wildness, the pleasure of muscles moving in coordination, the sweat and the weariness, and the uncertainty of what the end to that effort will be.

The pretrip physical conditioning, or the constant maintenance of it, must improve each year to offset the aging process. When I was forty-one I could get by on youthful vigor. When I am seventy-one I shall have to be able to do seventy-one pushups. *Vive moi, viva me.*

Always I come back from these trips feeling like a skinned-up kid, feeling like a renewed, recreated adult, feeling like a tiger. All that basic nature, all that use of animal instincts, arouses some very earthy desires. The most delicious comment about these trips was by a sailor-oceanographer who understood the sea both mentally and physically. "A woman who would do a thing like that," he said, "is worth going to bed with." A classic remark—but it said more about him than about me. You don't make love to just a female body. It doesn't matter if she's twenty

130

or eighty; you want the whole person. At least that's the best kind, and at least I thought that's what it said about him.

I paddle on. Through the rocky pass at the point of Kūka'iwa'a, on toward the high sea stack of Huelo. The fan shaped leaves of the *loulu* palm at its crest glisten in the sun. The old Hawaiians had a game of attaching the leaves like wings and leaping off into the water—the first hang gliders. I paddle completely around it and look back past its tower to the peak of Hā'upu.

Hoe aku i ka wa'a. I chant the old paddling rhythm.

The future of this coast is uncertain. Population pressures bring more people to a wilderness, more boats, more hikers, more young adventurers who paddle surfboards, hike, and swim. There is increasing interest in the water resources of the upper valleys to provide for the planned Kaluakoi resort development at the west end of the island. Tunnels may siphon off most of the water from the rivers. Like the island of Kaua'i and its roadless north coast, there may be the daily helicopters and the summer boats bringing the loads of tourists and campers and hunters who leave their trash on the wild shores and then zoom off. There should be some areas everywhere in the world where motorized access is forbidden, places you get to only through the natural, quiet energy of wind and muscle.

North Moloka'i is only a twenty-mile coast; it is a small fragile area and will not survive an onslaught of people. We have not yet learned that all Hawai'i is a small, special place that we shall soon destroy with deliberate, carefully organized "progress." We do not yet know how to care for it.

And why did I always come alone to Moloka'i? I know why, but the telling is hard. Daily we are on trial, to do a job, to make a marriage good, to find depth, serenity, and meaning in a complex, deteriorating world of politics, false values, and trivia. But rarely are we deeply challenged physically or alone. We rely on friends, on family, on a committee, on community agencies outside ourselves. To have actual survival, living or dying, depend on our own ingenuity, skill, or stamina—this is a core question we seldom face. We rarely find out if we like having only our own mind as company for days or weeks at a time. How many people have ever been totally isolated, ten miles from the nearest other human, for even two days?

Alone, you are more aware of surroundings, wary as an animal to danger, limp and relaxed when the sun, the brown earth, or the deep grass say, "Rest now." Alone, you stand at night, alert, poised, hearing through ears and open mouth and fingertips. Alone, you do not worry whether someone else is tired or hungry or needing. You push yourself hard or quit for the day, reveling in the luxury of solitude. And being unconcerned with human needs, you become as a fish, a boulder, a tree—a part of the world around you.

I stood once in midstream, balanced on a rock. A scarlet leaf fluttered, spiraled down. I watched it, became a wind-blown leaf, swayed, fell into the water with a giant human splash, then soddenly crawled out, laughing uproariously.

The process of daily living is often intense and whimsical. The joy of it, and the compassion, we can share, but in pain we are ultimately alone. The only real antidote is inside. The only real security is not insurance or money or a job, not a house and furniture paid for, or a retirement fund, and never is it another person. It is the skill and humor and courage within, the ability to build your own fires and find your own peace.

On a solo trip you may discover these, or try to build them, and life becomes simple and deeply satisfying. The confidence and strength remain and are brought back and applied to the rest of your life.

I go on to Waikolu, to Father Damien's church, and to the plane at Kalaupapa, on to people and problems and coping. But it is all right. Always now, Moloka'i is there, it is part of me. I can return to the lonely splendor, and I am no longer afraid.

EQUIPMENT *

Clothing

high topped tennis shoes
rubber slippers (zori)
socks
jeans
jacket
shirt
T-shirt
underwear
hat

Personal

towel or diaper
detergent
olive oil
moisturizer
toothbrush
comb
mirror
lipstick
scissors
Band-Aids
aspirin
needle

Miscellany

notebook and pencil
hand lens
binoculars
hammock
Mallory flashlight
waterproof flashlight
extra batteries and bulbs
books
candles
extra matches
reading glasses
flares
fishing gear
permit for Kalaupapa

Photography

Nikonos camera
Konica camera
film
clamp tripod
light meter and housing
flash bulbs and attachment
waterproof bag

* This is the deluxe list. Not everything is essential.

Transportation

backpack
boat
paddle
screwdriver
extra screws
extra paddle blade
topographical map
tide chart
pump
patch kit
finsmaskandsnorkel
gloves
plane ticket

Cooking and Food

pot and lid
bowl
cup
spoon
knife
grill
nylon net
matches
water bag
stove and fuel
food, bagged
wine

Shelter and Sleeping

plastic sheet
sleeping bag
air mattress
nylon cord
ground sheet

919.69
S

Sutherland, Audrey,
1921-

Paddling my own
canoe

919.69
S

Sutherland, Audrey
Paddling my own
canoe.

DEC 82   N P 4 3 4 0 7
DEC 82   N P 4 3 4 0 7

6 MAR 79   A P 5 1 4 0 2
19 MAR 79   A P 2 4 6 3 3
10 APR 79   N P 3 6 9 9 1
4 MAY 79   A P 2 9 7 6 6
1 JUN 79   N P 3 0 6 9 4

18 AUG 82   N P 3 5 2 5 7

2/79

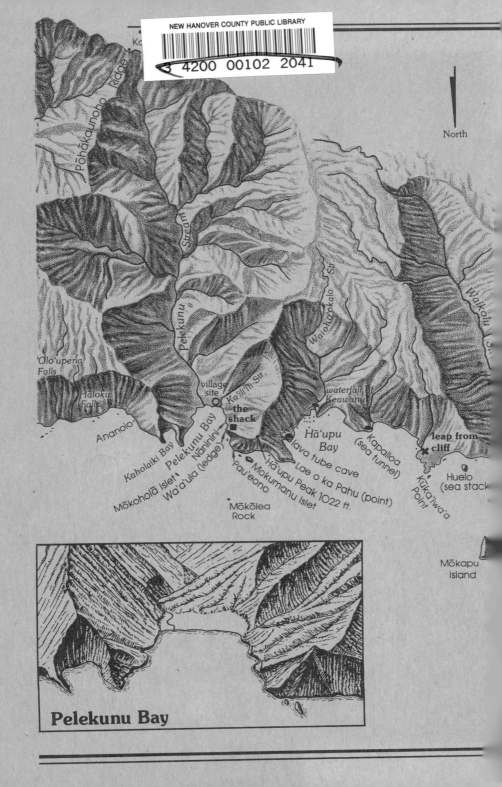

North

Pōhākaunoho Ridge

Ko

Pelekunu Stream

Waiohaʻokalo Str.

Waikolu Str.

'Oloʻupena Falls

Haloku Falls

village site

Kaʻiliʻili Str.

the shack

waterfall Keawanui

leap from cliff

Ananoio

Kaholaiki Bay

Pelekunu Bay

Nāninini

Waʻaʻula (ledge)

Mōkoholā Islet

Pauʻeono

Haʻupu Bay

lava tube cave

Lae o ka Pahu (point)

Haʻupu Peak 1022 ft.

Mokumanu Islet

Kapailoa (sea tunnel)

Kūkaʻiwaʻa Point

Huelo (sea stack

Mōkōlea Rock

Mōkapu Island

Pelekunu Bay